CLOUD STUDIES

LECTOR HOUSE PUBLIC DOMAIN WORKS

CLOUD STUDIES

ARTHUR WILLIAM CLAYDEN

ISBN: 978-93-5344-935-3

Published: 1905

LECTOR HOUSE LLP
E-MAIL: lectorpublishing@gmail.com

A SUNSET SKY.

CLOUD STUDIES

BY

ARTHUR W. CLAYDEN, M.A.

PRINCIPAL OF THE
ROYAL ALBERT MEMORIAL COLLEGE, EXETER

1905

PREFACE

To the meteorologist I hope the following pages may prove not only of some interest, but of practical value as a small step towards that greater exactness of language which is essential before we can attempt to explain all the details of cloud structure, or even interchange our ideas and observations with adequate precision. The varieties depicted and described have been selected from many hundreds, as those which seem to me to show such differences of form as to imply distinct differences in the conditions to which they are due. I have not attempted to deal with the physical causes of condensation except in a general way, being unwilling to introduce diagrams of isothermals and adiabatics and such purely scientific methods into a work also intended for a wider public. For those who wish to pursue this part of the subject I have appended a list of papers from the *Quarterly Journal of the Royal Meteorological Society* and other sources, which may serve as references. I also hope that some more votaries of the science may be induced to realize that meteorology does not consist solely of the tabulation of long columns of records, but includes subjects for investigation as much more beautiful as they are more difficult.

To the artist I trust they may also be of some use, by calling attention to the variety and exquisite beauty of the sky. Nothing is more extraordinary in art than the general negligence of cloud-forms. Many of them are quite as worthy of careful drawing as the leaves of a tree, the flowers of a field, the ripples on a stream, or the texture of a carpet, or a marble pavement. Yet it is the common rule to find pictures, which are otherwise marvellous examples of skill and care, disfigured by impossible skies with vague, shapeless clouds, as untrue to nature as it would be possible to make them. Grace of outline, delicacy of detail and texture, richness of contrast, beauty of form and light and colour, all are present in the skies, and combine to make a whole well worthy of the best that art can give. The illustrations I offer are not selected for pictorial effect; they are chosen from a purely scientific point of view; but they are enough to indicate what could be done if the facts of nature were treated with high artistic skill.

In addition to the meteorologist and the artist, there are a much larger number who follow neither profession, but who love Nature in all her moods; and to them also I hope these pages may be of interest. Indeed, if only a few of them should be stimulated to take up a branch of nature study which has given me many an hour of quiet enjoyment, the labour of bringing these notes together will not have been in vain.

ARTHUR W. CLAYDEN.

St. John's,
Exeter.

CONTENTS

LIST OF ILLUSTRATIONS

CLOUD STUDIES

CHAPTER I

INTRODUCTORY

ALL who have the faculties proper to man must have been to some extent students of cloud form. Go where we will, do what we will, we cannot easily escape from the sky, or avoid noticing some of its features and coupling them with the varying conditions of weather. We all sometimes want to know if it is likely to rain, or whether some other change is probable; and experience soon shows us that the clouds give the simplest and most obvious indication of what we may expect. It is almost impossible to avoid noticing that certain types of cloud, or the simultaneous appearance of certain types, is the usual accompaniment of definite kinds of weather or of particular changes. Thus it is that most people acquire some small measure of weather wisdom before their schooldays are over.

Generation after generation, through all human history, the same causes must have led to the same conclusions; and the study of clouds must, therefore, be one of the oldest of all branches of scientific inquiry. Yet, old as it is, it is still in its infancy, having made very little advance indeed towards the precision of an exact science.

There are many reasons for this want of growth, and so far as the theoretical aspects of the subject are concerned it is easy enough to understand. Clouds are among the most inaccessible of terrestrial objects. Except by balloon ascents, by sending up kites bearing recording instruments, or by making observations among the mountain-tops, we have no means of getting at them to study the conditions under which they exist. Temperature, pressure, humidity, have generally to be guessed at, those guesses being based on the scanty data which have been laboriously obtained by one or another of these cumbrous methods. Moreover, many clouds have such vast dimensions that it is very difficult to grasp all that goes on in such a space.

Besides the difficulty of attacking the problems presented by cloud formation, it is probable that even if we could have got among the clouds at will, we should have understood little more than we do, from a want of sufficient certainty on many of the purely physical questions involved. It is not many years since Mr. J. Aitken discovered the necessity for material nuclei as a first step in the formation of cloud particles, and not many months have elapsed since Mr. C. T. R. Wilson showed that those particles can be formed by the action of radiation on the air itself. There is nothing surprising, therefore, in the fact that our theoretical knowledge of the why and wherefore of the facts revealed by a study of clouds is limited to general principles, and quite fails to say exactly why each special form should be assumed. The matter for surprise is quite different.

Theoretical explanations are not the first step in the working out of a branch of science. It begins with the acquisition, by diligent and painstaking observation, of a great mass of facts. This may go on for centuries, the accumulation growing greater and greater, until at last some one comes who examines the records, classifies them carefully, and finally makes a summary in the form of a number of generalizations, which are announced under the name of Laws.

Two examples of such "Laws" will suffice. Astronomers for centuries had observed the movements of the planets, always with increasing accuracy, until Tycho Brahe made his famous series of observations on the planet Mars. These materials fell into the hands of Kepler, and the result of his work was the announcement of Kepler's Laws, which state the rules which govern the movements of the planets in their orbits. He found that the records could not be accounted for unless the planets moved in a certain way, but he knew nothing of the reasons for a method and order which clearly existed.

Kepler's Laws, in fact, rest upon another set, namely, Newton's Laws of Gravitation, and these are themselves a second example. They are the summary of accumulated experience, and even at the present day we know nothing certain as to why two bodies should attract each other, and nothing as to why that mutual attraction should act as it was found to act by Newton.

The observational part of cloud study, however, is still in its infancy, in spite of the fact that it has been going on for such countless ages. We are still in the condition of the humble observers engaged in the comparatively humdrum task of gathering facts for future arrangement and interpretation. Cloud observers, in all ages, have suffered from a peculiar difficulty. They have had no common language, no code of signs by which they could benefit from the work of those who had gone before them, no means of transmitting their own experience to each other, or to those who would come after them. No progress would be possible in any study under such conditions, for each person would begin where the previous generation began, instead of taking up the task where others had left it. In all languages there is an extraordinary scarcity of cloud names, and such as do exist are frequently applied to quite different forms by different people. So pronounced is this lack of terms, that any one who tries to describe a sky without using any of the modern scientific names, finds himself obliged to rely on long detailed descriptions, backed with references to well-known objects, whose outlines or structures resemble the clouds more or less vaguely; and even then he has to be a word-painter of singular skill if his description calls up in the mind of the reader a picture much like the original.

It was to meet this want of a common tongue that Luke Howard, in 1803, proposed his scheme of cloud names. He recognized three main types of cloud architecture, which he named Cirrus, Stratus, and Cumulus. Cirrus included all forms which are built up of delicate threads, like the fibres in a fragment of wool; Stratus was applied to all clouds which lie in level sheets; and Cumulus was the lumpy form.

By combinations of these terms other clouds were described. Thus, a quantity of cirrus arranged in a sheet was called cirro-stratus, while high, thin clouds like cirrus, but made up of detached rounded balls, was cirro-cumulus. Many cumulus clouds, arranged in a sheet with little space between them, became cumulo-stratus, while the great clouds from which our heavy rains descend partake, to some extent, of all three types, and were therefore distinguished by a special name—Nimbus.

This system had much to recommend it. The three fundamental types were obvious to all. Their names were descriptive, and were derived from a dead language, so that no living international jealousies were raised. It was sufficiently detailed to serve the purposes of the time, when accurate observation was in its infancy. Hence it was universally adopted, and will pretty certainly hold its own as the broad basis upon which any more detailed system must necessarily rest.

It has done excellent service; but although observation of clouds in a general way is far from complete, attention is now being given to much smaller details and much more minute differences of form, and our vocabulary must be amplified. Precision of description is the first essential of a satisfactory system, and the question is, what sort of edifice can we build on Luke Howard's foundation.

The great difficulty is the infinite variety of clouds. Certain forms may be arbitrarily selected as types, and names may be given to them; but however well they are chosen, a very short period of observation will show that there are all manner of intermediate forms, which make a perfect gradation from one type to another. This fact should never be forgotten. There is always a danger that the use of any system of names based on types shall lead to the neglect of everything not typical. A curious illustration is afforded by the well-known fact, that in arranging collections of fossil shells, it is frequently found that some specimens do not exactly match the type examples to which names have been assigned. In former days it was the custom to throw aside such "bad specimens," as they did not show plainly the specific characters. It is now realized that they have a value of their own, in that they are the links in the evolutionary chain, once supposed to be missing. Indeed, it is not unfrequent nowadays to see carefully selected series, showing the gradual change whereby one species passed into another, displayed in the place of honour, while the type specimens are relegated to humbler places in the general collection.

Types there must be, no doubt, and where the series is continuous, some one must make the selection. With clouds the series is absolutely continuous. The task is like choosing typical links from a long chain in which each link is almost exactly like its neighbours, yet no two are alike, and the greater the distance between them the less their likeness. Clearly any system put forward must be accompanied by illustrations, so that all may know exactly which links have been chosen.

Many attempts have been made to meet the want; some of the systems proposed being based on the forms assumed by the clouds, some on their supposed mode of origin, and some on their altitudes. Those which were not founded on Luke Howard's types had no chance of being accepted, while knowledge was not yet sufficiently far advanced to make classifications based on origin of form at all possible. But the great reason why none of the proposed schemes could come into general use was that they were put forward without adequate illustration, so that none but their authors knew exactly what they meant.[1]

Matters came to a head in 1891, when an International Meteorological Conference met at Munich. One object of this gathering was to promote inquiries into the forms and motions of clouds, by means of concerted observations at the various institutes and observatories of the globe. Luke Howard's system was not enough for the purpose in view, and the addition of more detailed terms had to be settled before work could be begun.

Professor Hildebrandsson, of Upsala, and the Hon. Ralph Abercromby jointly submitted a revised scheme, the main feature of which was the introduction of a new class of clouds, to be distinguished by the prefix alto-before the other name. Such alto clouds were less lofty and denser than cirrus. This scheme was the best before the Conference, and without waiting to discuss, and possibly improve it, it was formally adopted, and a committee appointed to arrange and publish an atlas showing pictures of the type-forms. This atlas did not appear until 1896, and in the mean time the Rev. W. Clement Ley had published proposals of his own, some of which had much to recommend them. But he was too late. The International Committee had come to a decision, and, although it may be far from ideal, the system backed by such an authority must be regarded as the standard until some similar gathering has given worldwide sanction to a change, and even then it would be better to modify by addition rather than by substitution.

The subjects of the following pages are named in general accordance with this International Code, but they are by no means restricted to types. Their object is not to attempt any repetition of the work which has already been well done by the Atlas Committee, but rather to show the chief varieties within a type. It will, however, become abundantly evident that the standard system is far from complete, and that any minute and detailed study of cloud-form must take note of the precise variety.

This at once raises the question whether many of these varieties are not sufficiently distinct to be given definite names. If a meteorologist is told that cirrus clouds were seen on a particular occasion, he instinctively asks—What sort of cirrus? and is utterly unable to form any mental picture of the clouds until the question has been answered by a detailed description. A glance at a few of the plates further on will show the difficulty plainly, and it occurs with other forms of cloud as well as cirrus.

Is it not time that the International names were regarded as those of the cloud genera, and to add specific names for those varieties which seem to imply some difference in kind in the conditions which have led to their formation? This has been here attempted by translating into Latin the ordinary English term by which the variety would naturally be described. More extended observation will probably show that other species should be introduced, and possibly some of those

---

[1] See reference No. 2 on p. 109.

suggested in these pages may have to be subdivided. Whatever the names may be, specific distinction of some sort is an essential preliminary to detailed study of the why and wherefore of the particular forms.

The International system is as follows:—

A. Upper clouds.

>   (*a*) Cirrus.
>   (*b*) Cirro-stratus.

B. Intermediate clouds.

>   (*a*) Cirro-cumulus and alto-cumulus.
>   (*b*) Alto-stratus.

C. Lower clouds.

>   (*a*) Strato-cumulus.
>   (*b*) Nimbus.

D. Clouds of diurnal ascending currents.

>   (*a*) Cumulus and cumulo-nimbus.

E. High fogs.

>   (*b*) Stratus.

In this tabulation the forms marked (*a*) are detached and occur in dry weather, while those marked (*b*) are widely extended. The original scheme also gives the mean heights of the various types, but these values have been omitted here because they are extremely variable, and impossible to ascertain with any approach to accuracy by mere eye estimates. They vary also with the season, and probably also with the locality. Moreover, the altitude is no guide to the name, except that on the whole the types occur in the order given, taking group A as the highest and group E as the lowest. In the chapter on cloud altitudes this subject will be further considered, and under the descriptions of cloud-forms their average height or actual measurements for the particular specimen figured will be given whenever possible.

Before coming to the description of individual forms, it may not be out of place to give brief consideration to the best means of observing them in nature. For eye observation, of course, no directions are needed when we are dealing with the lower and denser varieties; but when we come to the highest groups it sometimes becomes necessary to protect the eye from the brilliant glare which may make it impossible to detect the real structure. Smoked glass, neutral-tinted spectacles, or yellow glass all have something to recommend them; but by far the most convenient means is to look, not at the clouds themselves, but at their images formed in a black mirror. A lantern cover glass, or a thin piece of plate-glass, blacked on the back with some black paint, serves admirably. But all black paints are not equally good. The best are oil paints which dry with a glossy surface, the so-called enamels. They have the advantage that the varnish with which they are mixed has an index of refraction not very different from that of the glass. The consequence is that so little light is reflected from the blackened back, compared with that which is reflected from the front surface of the glass, that the second image can only be detected with difficulty. If the mirror is a piece of black or deeply coloured glass all trace of the second image is lost.

With this simple appliance it is easy to study the details of the thinnest clouds right up to the sun, and even the image of the sun itself may be glanced at without serious discomfort. Nor is the general diminution of brightness the only gain. If the glass is so held that the light from the cloud makes an angle of about 33 degrees with the surface, some of the blue light from the sky is suppressed altogether, while that from the cloud is practically unaffected. The exact fraction suppressed depends upon the part of the sky relative to the sun, and also on the position of the mirror, but a few minutes' trial will show when the maximum effect has been reached.

It is astonishing to see for the first time how the delicate filaments of cirrus or the beautiful structures of cirro-cumulus stand out shining white on the deep blue background; and the use of the black mirror is a revelation to most. It also has one indirect advantage, which is really more important than it seems. By gazing down into a mirror long-continued observations can be made, and one form of cloud may be watched changing into another, and possibly back again into its original shape, without any danger of incurring that unpleasant result of much looking upwards which is sometimes known as exhibition headache. Such a mirror may be quite small, so that it can be carried in a pocket-book, a point of some moment, as many of the forms of cirrus are exceedingly transient, coming and going in a few minutes, while others are in a state of continuous change. This is particularly often the case with the exquisite ripple clouds, and the delicate lacework of the higher kinds of cirrus.

Still another advantage possessed by the mirror is that it makes it easy to see the solar halos formed on the verge of a cyclone, and to detect their iridescent colouring in a way which is quite beyond the reach of the naked eye or any protective spectacles. Every one is familiar with the faint halos formed round the moon, but the corresponding solar phenomenon is comparatively little known, though it is far commoner, much more brilliant, and often glows with colour. Its very brightness, and that of the background on which it is projected, hides it from the eye, except on those rare occasions when the sun is conveniently hidden by some thicker cloud.

If some permanent record is desired, much can be done with a few light strokes of a pencil, but more ambitious pictures are best secured by the use of soft pastels, aided by a liberal use of the finger or leather stump. Ordinary paints, whether oil or water-colour, are of little use for actual study of cloud detail, except in the hands of a highly skilled artist who knows how to get the effect he wants in the minimum of time.

But no sketching or drawing can make records of cirrus or alto clouds with the speed and accuracy necessary for careful study. Photography is really the only way in which the amazing wealth of detail can be truthfully portrayed. Yet even the camera has its limitations. It does not record colour, and completely fails to delineate the forms of alto-stratus, stratus, or nimbus, if they are present in the most typical condition, that is to say, when they cover the whole sky with a uniform tint. It is only when these forms are more or less broken up that a photograph, or anything other than a carefully coloured picture, will represent them at all.

Cloud photography, even of the most delicate and brilliant varieties, is easy enough when the right methods are followed; but these are not the same as those which are right for portraiture or landscape work of the usual kind. The background of blue sky produces almost the same effect on the plate as the image of the cloud itself, and the whole art consists in an adequate exaggeration of the minute difference so as to reveal the details of form and structure.

A slow plate—the accompanying illustrations have all been taken on Mawson and Swan's photo-mechanical plates—extremely cautious development, and sometimes intensification of the image, are all that is necessary; but the process becomes easier if, instead of pointing the camera to the cloud, it is directed to the image formed in a properly constructed black mirror. Many of the following studies have been taken by this method, and details of the camera and processes employed will be found in a later chapter, for the convenience of any one who may be inspired to take up a fascinating branch of photography.

It has been said that reference will be made to the average altitudes of the different types of cloud, and to the actual altitude of some of the varieties shown. The question will, no doubt, have occurred to some as to how those altitudes have been measured. The methods are all more or less complicated, involving rather laborious calculation. They generally depend upon simultaneous observations made from two stations at opposite ends of a measured base line. Sometimes the observations are made directly by pointing an instrument at each station to some agreed point of the cloud. It is obvious that the two directions must converge to this point. If the convergence is measured, the exact distance from either station can be calculated, and if the angle between the cloud-point and the horizon beneath it is noted, it is a simple matter to deduce the actual altitude of the cloud. At other places the observers have relied upon the comparison of photographs simultaneously taken from the two stations. In this method it is necessary to know the exact direction in which the camera is pointed, and the position of the image upon the plate then gives the direction of the cloud as seen from that particular station, and the subsequent calculations are the same.

Measurements by one or the other of the above methods have been made at several places, the most extensive series being those which have been compiled at Upsala, and at the Blue Hill Observatory in Massachusetts. The method employed by the writer at Exeter has been rather different, and a description will be found later on in the chapter on Cloud Altitudes, the fuller consideration of which comes naturally after the different forms have been described and compared.

CHAPTER II

CIRRUS

A CLOUD is sometimes defined as any visible mass composed of small particles of ice or water suspended in the air, and formed by condensation from the state of vapour. As a general rule this is exact enough, but under certain circumstances it is possible to have the particles so small, and so thinly scattered, that it is not fully satisfied. The resulting mass may not be actually visible. The presence of the condensed particles may be indicated by nothing more than a slight whitening of the blue sky, or by the formation about the sun or moon of bright circles of light known as halos. If such a halo appears, it is generally a phenomenon of brief duration. Sometimes the circle breaks and becomes incomplete by the passing away of the thin patch of cloud, sometimes the cloud increases in density until the rings are destroyed.

The thinnest variety of this halo-producing structure is quite invisible to the eye. It is so thin as to have no distinctly noticeable effect upon the colour of the sky, but the optical results of its presence may be very remarkable. Highly complicated systems of rings are sometimes produced, the rings, as a rule, falling into two groups. The commonest form has the sun (or moon) in the centre, and a circle of pale light at a distance of about 22 degrees. Larger rings are seen less frequently, which have an angular radius of about 46 degrees, and as a rule have the sun situated on the ring itself. In Plate 1 we have a part of such a great halo. The camera was directed towards the east, and tilted upwards at an angle of about 40 degrees. The sun was behind the camera, in the south-west, and the ring could be traced right up to it on either side.

At the same time the sun was surrounded partially by a halo of the more ordinary type, which was brightly coloured, making an effective contrast to the dull white of the greater ring. The phenomenon did not last more than half an hour, and the changes in its appearance coincided with a growing density of cloud. When first noticed the great ring was alone, and the sky was of a full blue, but a silvery film came gradually up from the south-west, and the smaller and brighter halo flashed out as the delicate curtain came near the sun. Slowly the cloud spread to the north-east, gathering density from the opposite point of the compass; and by the time the ordinary halo was at its best, the great white ring had completely vanished.

Plate 1.

PART OF A GREAT HALO.

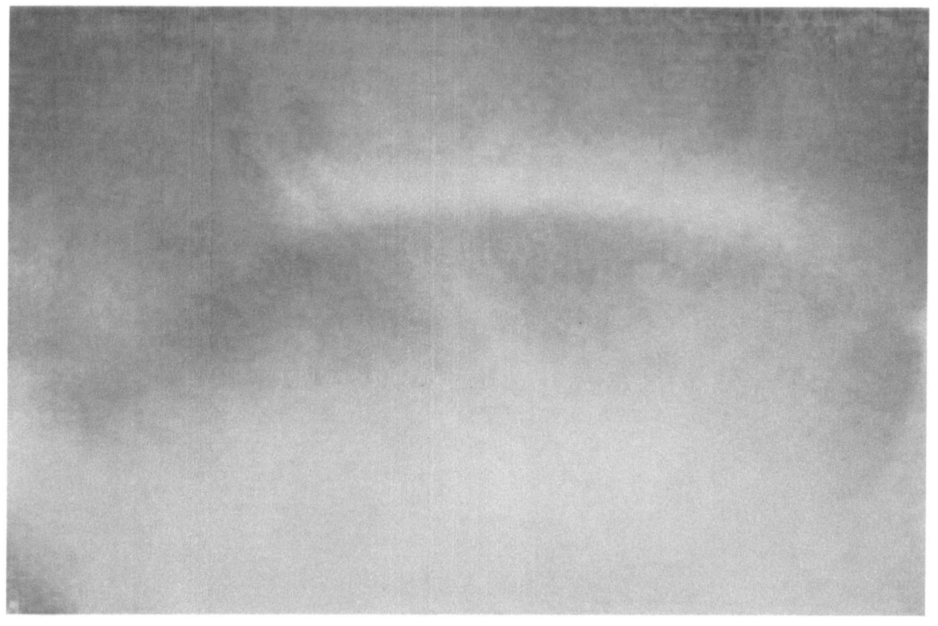

Plate 2.

PART OF A SOLAR HALO.

These circles, and the bright spots called mock-suns or mock-moons which often accompany them, can all be explained on the assumption that their cause is the passage of light through a veil composed of hexagonal crystals of ice. The simple halo of 22 degrees radius is common in most parts of the world, being very generally formed by the film of high cloud which marks the advancing edge of a cyclonic cloud system. A portion of one is shown in Plate 2, in which the rudimentary fibrous structure of the sheet of cloud is distinctly seen. Halos of this sort are frequently coloured, often most brilliantly so; but the tints are seldom noticed unless a black mirror is used. They are sometimes quite as bright as those of an ordinary rainbow, but instead of being projected upon a background of dark rain-clouds, they are seen against a part of the sky which is near the sun, and therefore exceptionally bright.

The red is always on the inside of the ring, the violet outside, thereby distinguishing them at once from the so-called coronæ, which are formed around the sun or moon when shining through a sheet of alto or other lower cloud made up of liquid particles. In these the radius of the rings is much less, and the red is on the outside, the violet actually touching the central luminary.

The cloud which produces halos is called cirro-nebula. It is much thinner, and on an average higher than cirro-stratus. Mr. Ley named it cirro-velum (or cirro-veil), but cirro-nebula has now got to be fairly well understood. It sometimes appears and disappears in a curious manner, showing that it occurs in patches, which drift about or which keep forming and melting away, only to repeat the process. If, however, it forms part of an advancing cyclone fringe, then the sky gets whiter and whiter, until it is covered with a sheet of undoubted cirro-stratus. This process of growing density is shown in progress in Plate 3.

Plate 3.

CIRRO-NEBULA CHANGING TO CIRRO-STRATUS.

Cirro-nebula, as we shall call it, floats at very great altitudes in temperate regions; but in polar latitudes, where the optical phenomena peculiar to it are most brilliant and diversified, it seems probable that the ice dust is much lower down, even in actual proximity to the ground. In England its height varies greatly with the time of year, and other circumstances, but mounts up in summer to such altitudes as nine miles or more; the greatest height yet recorded being 9·6 miles, or about 15,500 metres, at Exeter.

The change from cirro-nebula to cirro-stratus is generally accompanied by the formation of a distinct fibrous structure, easily observable by the black mirror. This is not really a new feature, but only a further development of a structure already existing, but too minute to be easily seen. True halo-producing cirro-nebula may usually be shown to possess more or less of a fibrous texture in an indirect way, which is worth a brief description.

In order to observe the spots on the sun and other features of the solar surface, it is a common practice to hold a white screen, say, about a foot from the eyepiece of a telescope, while the instrument is pointed to the sun. An image, considerably magnified, is thus projected on to the screen, and the solar details can be studied with ease and safety. If thin clouds drift before the sun, their images are similarly projected as they pass across its disc, and it is possible thus to detect not only the fibrous texture but also the movement of cirro-nebula.[2]

---

[2] The telescope with which these observations have been made is a 6·8-inch refractor

The change into cirro-stratus is also attended by a marked fall in altitude, but whether this is due to an actual descent of the cloud particles, or to a downward spread of the conditions which give rise to them, cannot at present be definitely settled. The balance of probability points very strongly towards the downward spread of the conditions. It is likely that the clouds, particularly the cyclonic specimens, are wedge-shaped, and that as they pass overhead we see first the thin edge, and later on the thicker parts, which project much lower down. This is just one of those many minor problems in cloud mechanics which we are not able to solve from the scanty data on record.

Occasionally cirro-nebula breaks up into little detached semi-transparent cloudlets, all of them exceedingly thin, and showing a complicated mottling, resembling, on a minute scale, the ripple clouds of much lower altitudes. Such a sky is depicted in Plate 4, but no reproduction can possibly do justice to the minute and delicate features of the real thing. The arrangement of the faint markings was in a state of continual flux, curiously similar to the ever-changing aspect of the sun's photosphere when seen under adequate power. Some parts of the cloud stratum would at one moment break up into distinct granules arranged in complicated patterns, other parts would assume a fibrous texture, and yet other places would show a continuous smooth sheet. In a minute or two all would be changed—the smooth part granulated, the fibres vanished, and the granules fused together, and so on, no two of a series of photographs representing the same details.

Plate 4.

CIRRO-NEBULA CHANGING TO CIRRO-CUMULUS.

equatorially mounted.

These changes of form continued until the whole was hidden from view by a veil of much lower stratiform cloud, one advance portion of which is shown. Plate 4 does not represent a type or a distinct variety of cloud. It is an intermediate form, or a temporary condition, showing cirro-nebula in the act of changing into cirro-cumulus, or possibly cirro-stratus.

Cirro-nebula itself, in its simpler form, is, however, a distinct type. It is true that it never persists over one locality for more than an hour or two without passing into some denser form, but while it lasts its features are so distinctive, and the optical phenomena to which it gives rise are so striking and significant, that it is a matter for surprise that it should in the International system have been relegated to the position of a subordinate variety of cirrus. It is more nearly related to cirro-stratus, but is sufficiently distinct from that to deserve at least specific rank.

True typical cirrus must have a plainly shown fibrous structure. The fibres may cross and interlace, they may radiate in fan-like manner, or they may curl and twist like a well-trimmed ostrich feather. The clouds so formed must not be arranged in a continuous level sheet, or they at once become cirro-stratus, and it is impossible to invent a definition which will mark the exact limits of either type. Typical cirrus consists of detached clouds. They cast no shadows on the landscape, for the simple reasons that they are semi-transparent and their component parts too narrow. If the sun is shining down obliquely through the naked boughs of a tall tree, it will be seen that the lowest twigs cast fairly sharp shadows on the ground, but that even these are bordered by a fading rim; the twigs further up cast no sharp shadows, but broader faint bands of shade; while the topmost boughs cast no shadows which can clearly be identified. In other words, the more distant the narrow twig is from the ground the narrower the real shadow or umbra, and the broader the penumbra becomes, until when the distance is sufficient the shadow is all penumbra. Cirrus filaments throw nothing but a faint penumbra. Indeed, it is only when they lie in the earth's shadow, and stand against the background of a faintly lighted sky, that they show any sign of shadow even on themselves.

There is no doubt that they are composed of particles of ice. They are formed at altitudes where the thermometer must be many degrees below freezing-point, and not a few of the thinner examples show fragmentary halos like those of cirro-nebula.

Their actual altitudes are very variable, being greater in summer than in winter, and reaching a maximum for any given station after a long spell of hot weather. Exact measurements have not yet been made in tropical latitudes or in polar regions, but there is every reason to expect that the upper limit of cirrus for equatorial districts will be found to be much higher than in the temperate zones where actual observations have been made. In places nearer to the Arctic Circle it is also almost certain that the altitudes will be less.

In the New England states, as shown by the Blue Hill observations, the maximum altitude for summer was found to be little under 15,000 metres. At Upsala, in Sweden, it was 13,300 metres. The average altitudes at the same observatories were, respectively, about 9900 and 8800 metres. At Exeter the writer's own measurements give an average for the summer months of 10,200 metres, with a minimum rather lower than was the case in America or Sweden, and with a maximum far above the foreign values. In winter cirrus certainly comes much lower down, but the number of observations is fewer.

Plate 5.

HIGH CIRRUS. (CIRRUS EXCELSUS.)

The loftiest variety of cirrus appears in the afternoon in very hot weather, sometimes quite late in the evening; and in autumn it is by no means a rare event for it to suddenly form just when the sunset colours are fading, or even after they have paled into twilight. Under such circumstances it stands out of a shining silvery grey colour against the background of the darkening sky. A specimen of it is shown in Plate 5, which shows its extreme slightness of form and delicacy of texture. Sometimes it remains visible so long after the stars have begun to show as to give the idea that it is self luminous, and the illusion is certainly very strong. The writer has noted several instances in which it was plainly visible, like a silvery curtain, though the sky as a whole was so dark that stars like the five brightest points of the Great Bear could be seen through the cloud, and much smaller stars down to the third and fourth magnitude were plainly visible in the clear intervals. It has sometimes been called luminous cloud, and Mr. Ley estimated its altitude at upwards of 90,000 metres; but if we think of it as reflecting the light of the distant colourless twilight there is no need to regard it as anything fundamentally different from other clouds, or to assume a greater altitude than we know to have been the case. The specimen figured occurred in the early afternoon on June 12, 1899, at Exeter, and careful measurements of its altitude were made. This worked out as 17·02 miles, or more than 27,000 metres, a value so much greater than all other measurements of the kind that it was only after most careful verification and reference to duplicate records that it could be accepted. It differs in several ways from the lower varieties, being thinner, more glistening, and in every way more delicate. A suitable distinctive name would be high cirrus, or cirrus excelsus.

Plate 6.

WINDY CIRRUS. (CIRRUS VENTOSUS.)

Lower down by thousands of metres come the feathery masses of typical windy cirrus, such as are shown in Plate 6. Indeed, in cold winter weather they occur within three or four thousand metres of the ground. In the instance figured the wind was blowing from left to right, and the clouds were travelling swiftly. The upper filaments appeared to be repeatedly torn away from the main masses, while the long faint streaks which trail below and behind are evidently due to streams of fine particles falling from the main centres of condensation into a less rapidly moving stratum below. There is no room for doubt that these clouds, like others of a similar order, are formed by a direct passage from the vapour to the solid, or that the fibres are made of minute snowflakes. The condensation is evidently attended by rapid movements, which draw out the cloud, as fast as it is formed, into long curving lines which mark lines of motion. The variety is always, therefore, an indication of strong winds and rapid eddying movements in the region in which it occurs. Such strong disturbances overhead almost always accompany similar but less intense movements at the ground-level, and when they do not accompany them they precede them. The cloud is well named windy cirrus, which may be converted into a specific name, cirrus ventosus.

The next variety we come to (Plate 7) is in some ways rather similar. It is, how-ever, thinner, more delicate, and is entirely composed of fine threads, which are more systematically arranged. Generally there is a bundle, or several bundles, of long parallel fibres, which form, so to say, the quill of the feather, with numbers of shorter threads branching out from them at various angles. Cirrus ventosus was indicative of irregular movements in various directions; this variety points also to complicated movements, but executed in accordance with some sort of system, strangely complex and wonderfully ordered. The specimen figured is the type of what Mr. Ley called cirro-filum, or thread cirrus, and his name can hardly be bettered. It is a cloud of summer, and occurs rather high up in the cirrus zone, but no actual measurements can be quoted. It is fairly common, but not nearly so frequent as the last.

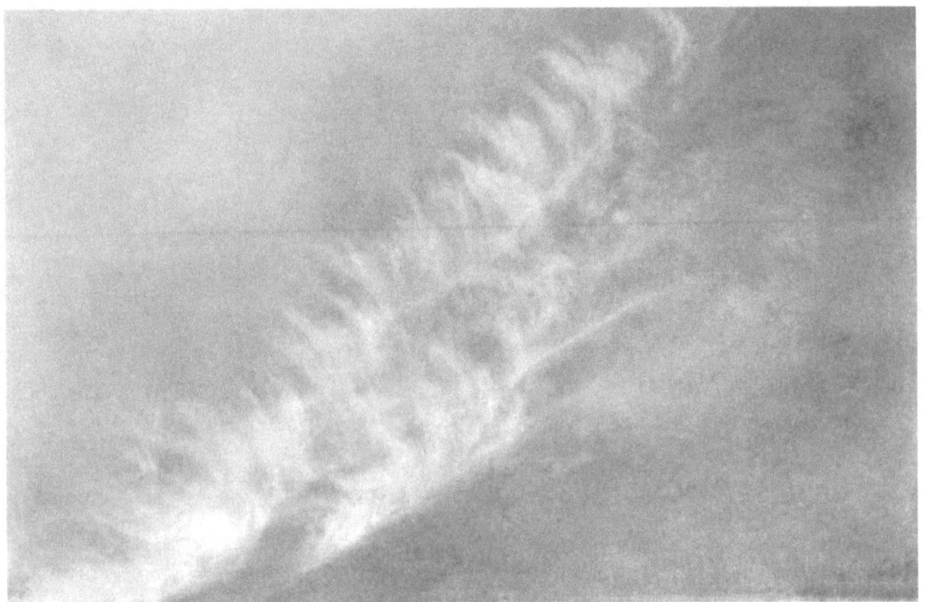

Plate 7.

THREAD CIRRUS. (CIRRO-FILUM.)

A somewhat more familiar variety is shown in Plate 8. Little irregular feath-ers of cirrus, from which long tapering streamers point downwards in graceful curves, or else lag behind in the direction from which the clouds have travelled. If clouds of this type are carefully watched, it will soon be seen that each feath-ery head is a centre of condensation, and that the tails or streamers are nothing else than falling particles, which dwindle slowly away by evaporation, and which gradually sink below the level of the heads. It is usual, in dealing with cloud-forms like these, to speak of air-currents of different velocities almost as if the winds at different levels were as clearly separated as oil and water, or even air and water. This can hardly be the case, for if such a thing should occur as an air-current of one velocity flowing over another of less speed, or of a current in one direction over another moving in a different course, the two must inevitably mix at their junction, and in a very short time the passage from the lower current to the upper one would be quite gradual. No doubt we can often observe two, three, or more layers of cloud moving in different directions; but if we were to send up a balloon, it would be rare indeed to find its direction of horizontal movement changed in a few metres of ascent. Different and distinct air-currents are often invoked to explain cloud-forms quite unnecessarily. It is far more likely that the differential movements involved in the explanation of the features of these cirrus varieties are due to increased velocity with greater altitude, to progressive change of direction, to irregular eddies, or to the interaction of ascending and descending convection currents. Indeed, it is probable that careful study of the growth and decay of these clouds will, in time, lead to a clearer understanding of atmospheric movements, and so enable us to say more precisely why they are as we see them to be. The variety shown in Plate 8 is rare except in combination with other forms. It might well be termed tailed cirrus or cirrus caudatus.

Plate 8.

TAILED CIRRUS. (CIRRUS CAUDATUS.)

The form of cirrus shown in Plate 9 is far more frequently seen than either of those which have been described. In this the fibrous texture is very imperfect, and the cloudlets show a tendency to arrange themselves in a kind of ribbed structure in two directions almost at right angles to each other. But this last is an accidental feature of the particular example, and not in any way a specific character of the cloud. The reason for regarding it as a distinct variety is the total absence of sharply defined lines, not only the heads of condensation, but even the long streamers attached to them being uniformly hazy and ill-defined. It is a form of cirrus which comes at all seasons, but most frequently in summer; it moves always with great slowness, indicating a quiet atmosphere free from disturbance of any kind. The conditions necessary for its appearance are a nearly uniform distribution of pressure over a considerable area, chequered by little shallow depressions of some trifling fraction of an inch. In hot weather these are the conditions under which thunder-storms develop, and this hazy cirrus, or cirrus nebulosus, may be taken as a certain sign of such an atmospheric state.

Plate 9.

HAZY CIRRUS. (CIRRUS NEBULOSUS.)

So far as permanency of form is considered, hazy cirrus is one of the most persistent, and affords a marked contrast to the species shown in Plate 10, which represents the most fugitive. Five minutes before the photograph was taken the same part of the sky was a deep, clear blue, without any trace of cloud. Suddenly a few short curling wisps made their appearance. These rapidly increased in number, until a delicate filmy network extended over the greater part of the field of view. But while the camera was being adjusted for an exposure, part of the net had broken up into the granular structure shown in the lower part of the photograph. The granulation rapidly spread through the net, almost as if the fibres had been curdled, and five minutes later the whole had been converted into a patch of cirro-cumulus which soon fused into a uniform sheet. Meanwhile the same series of phenomena were taking place in other parts of the sky.

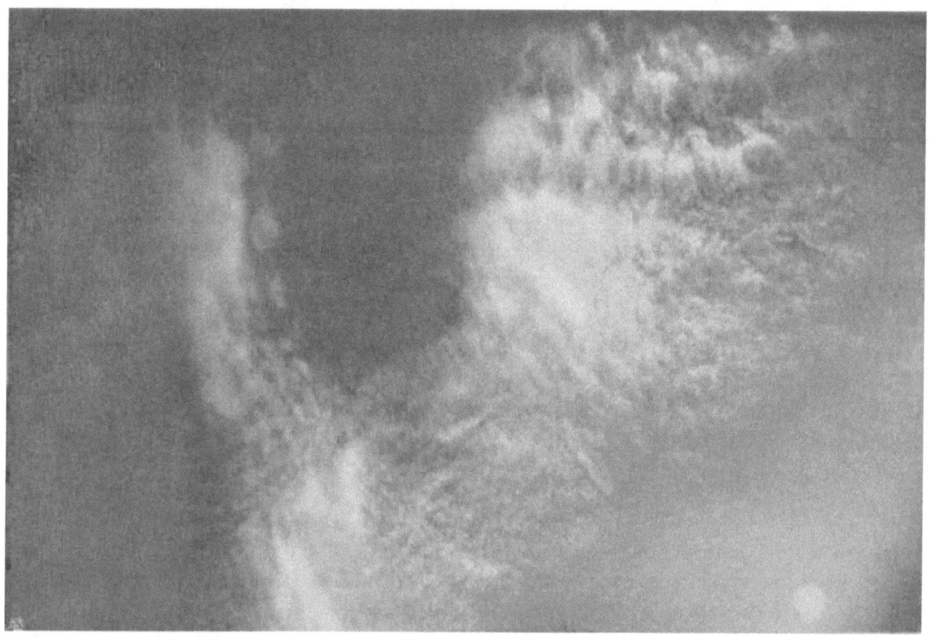

Plate 10.

CHANGE CIRRUS. (CIRRUS INCONSTANS.)

On other occasions exactly the same set of events have been seen to follow each other in the inverse order. Beginning with a fairly even sheet, this broke up into granules, and they in turn seemed to be frayed out into short hazy and wavy fibres which slowly melted away.

Clearly we have here to do, not with a distinct type of cloud, but rather with the first step towards the formation of one, or the last stage in the life of one which is drying up. But sometimes the life of the cloud is so short that it never passes beyond this first stage; and it is by no means a universal rule for a growing sheet of cirrus to pass through this stage at all. It therefore represents a peculiar state of instability, and requires a name of its own. Sometimes patches of it will come and go in an apparently capricious manner for an hour or more before permanent condensation is effected or before the sky finally clears. But this is a rare event, since the slow change of conditions which has brought the stratum of air to the unstable condition is generally progressive, and instead of stopping at the critical point, goes beyond it, with the result that the condensation grows or the cloud disappears entirely. Change cirrus, or cirrus inconstans, would be an appropriate name for a kind of cloud which is so plainly indicative of instability.

The critical condition referred to is, of course, that in which a particular stratum of air is just saturated, or is just on the point of forming visible cloud. If any cause is brought to bear on such a stratum which brings about even slight cooling, cloud must be produced; and, conversely, anything which results in the slightest heating must cause it to disappear. The shortness and haziness of the fibres, and the fact that they gather themselves into granules, shows that the cloud is formed in a stratum of air which is either still, or is moving as a whole, without any of those differential movements which seem to be necessary for the longer fibrous details.

The causes which may bring about the local cooling and heating are easy to understand when we remember how the air will be affected by the uneven contours of the ground. As it passes over hill and valley the up-and-down movements of the lower layers, or even the disturbances caused by passing over a wood or clump of trees, all must be propagated upwards. Each disturbance must slowly spread laterally and diminish vertically, so that it will reach the cirrus zone as a broad and gentle dome-like oscillation. Suppose now a series of such slight upheavals to reach the critical level. The passage of the waves will mean alternate expansion and compression. Expansion means cooling, and therefore cloud-production; compression means heating, and therefore the destruction of cloud.

From the most transient form of cirrus we pass, in Plate 11, to the most persistent and probably the most frequent. It occurs in detached masses which have very variable forms but are wholly fibrous, with the details arranged in a very irregular manner. The example figured was taken in the evening during a long spell of fine weather. If such a cloud is watched, its permanence of detail is very striking, and must be due to a persistence of slow eddying movements and to a continual renewal and waste of the component particles of each wisp. This is the kind of cirrus selected generally as the type of cirrus, and the selection is a good one. Common cirrus, or cirrus communis, it should be called. Settled conditions and fine weather are its usual attendants.

Plate 11.

COMMON CIRRUS. (CIRRUS COMMUNIS.)

We next come to a variety which is anything but a harbinger of good, namely, the long stripes or bands of cirrus which stretch outwards from the margin of the cloud canopy of a cyclonic storm. In some ways these appendages to the great nimbus resemble the strips of cirriform cloud which fringe the summit of a thunder-cloud. They look as if they must have been formed by the blowing away, by a rapid wind, of the top of an uprising column of vapour-charged air. Their main outline may thus be easily accounted for, but we have only to study their detailed structure for a few minutes to feel that they really present a problem of a very high order. Plate 12 shows a fairly simple example, but Plate 13 represents a cloud of very great complexity. To take this last the camera was tilted upwards at an angle of 45 degrees, so that the top of the picture is not far from the zenith. The wonderful plume of cloud rose from the southern horizon, and ended in a great sheaf of fibres and films spread out like a partly opened fan whose edge was only about 50 degrees above the northern horizon. Its length as it passed overhead lay between a point a little east of south to a little west of north; and the broad band moved as a whole, without any marked internal changes, from the south-west towards the north-east. The weather was very unsettled. A long procession of cyclones had been sweeping along our western shores, and the barometer was just beginning a fresh and rapid fall. During the ensuing night a heavy gale burst over the south of England.

Plate 12.
BAND CIRRUS. (CIRRUS VITTATUS.)

Plate 13.

BAND CIRRUS. (CIRRUS VITTATUS.)

The whole phenomenon was highly characteristic. These great bands with the divergent striation might well be known as storm bands, from their almost invariable connection with the violent atmospheric movements to which they are most probably due. Plate 12 shows a much less dangerous variety of the same species, which is distinguished from it by the comparative absence of internal detail and by the curled ends. Clouds of this character have sometimes been called cirro-filum, but a comparison of the plates with the typical cirro-filum of Plate 7 will show that there is little resemblance; and the attendant weather is also in marked contrast, both of which facts imply a fundamental difference in the causes to which their features are due. Banded or ribboned cirrus is the name which they immediately suggest, and this may be rendered cirrus vittatus.

This ends our survey of cirrus clouds. Any one who compares the plates so far given will see that they represent forms so diverse that it is impossible to avoid the conclusion that the conditions under which they are produced must differ not only in degree but also in kind. What those conditions are we have attempted here and there to suggest, but in no case can we feel that the explanation has been at all complete. In some cases, notably the last, we are face to face with such complicated details that it is hopeless to attempt to explain them in the present state of our knowledge. Fact upon fact must be accumulated until we can give their history from their earliest beginnings; and far more accurate and detailed knowledge of the attendant atmospheric conditions must be acquired before we can hope to rob such elaborate structures of their present mystery.

This requires the co-operation of many eyes and many minds, and exact specific names must be an essential preliminary. Those which have been suggested in these pages are—

1. Cirro-nebula, or Cirrus haze.
2. Cirrus excelsus, or High cirrus.
3. Cirrus ventosus, or Windy cirrus.
4. Cirro-filum, or Thread cirrus.
5. Cirrus nebulosus, or Hazy cirrus.
6. Cirrus caudatus, or Tailed cirrus.
7. Cirrus vittatus, or Band cirrus.
8. Cirrus inconstans, or Change cirrus.
9. Cirrus communis, or Common cirrus.

CHAPTER III

CIRRO-STRATUS AND CIRRO-CUMULUS

SEVERAL of the varieties of cirrus already discussed may gather so abundantly at some given level in the atmosphere, that the most obvious feature comes to be this arrangement in a sheet. The cloud then becomes cirro-stratus, and should be so named. We have described how cirro-nebula frequently grows in density until it fails to produce halo phenomena, and may even reduce the sun to a hazy patch of light showing no outline. This is the most typical of all forms of cirro-stratus. It has always a distinctly fibrous or streaky appearance, whereby it is at once distinguished from a similar but lower cloud which will be described later on.

A similar sheet may be formed from the fusion together of the streaks of cirrus-nebulosus, the bands of cirrus vittatus, or the development of cirrus inconstans. But the general rule is that the cirro-stratus retains more or less of the specific characters of the parent form.

Plate 14 shows a hazy form of cirro-stratus developed from the nebulous cirrus. Its altitude was great, being about 10,000 metres. The processes of growth and change could be studied easily. First would appear some faint spots and streaks; these quickly fused together into larger patches, which again joined to their neighbours. In a few minutes the cloud so formed would return to the mottled or streaky appearance, and either disappear entirely or become very thin, only to recommence the process. This went on for more than an hour, the cloudy patches getting larger and larger, until the critical condition was passed, and the sky was covered with a general veil of typical cirro-stratus.

Plate 14.

HAZY CIRRO-STRATUS. (CIRRO-STRATUS NEBULOSUS.)

In Plate 15 we have an example of a cloud which is clearly cirro-stratus, but the sheet is broken up into long bands, and each of these is made up of common cirrus. In the upper part of the picture the sprays of cirrus are forming, and as they come into being they are arranged in rows. We have here to do with a phenomenon of a very different order from the one presented by the true banded cirrus. The arrangement into belts must here be due to some kind of wave-movement in the air, breaking up the critical plane into long ribs transverse to the direction of the wave-movement. The specimen shown was moving in a direction nearly at right angles to the bands, though the surface wind was nearly parallel to their length. The question at once presented itself as to whether the movement of the bands was really a drift of the cloud, or whether it was not a case of the propagation of cloud production with the advancing wave. This was easily answered by watching the details of a band. The advancing side was always feathery, and careful observation showed that the edge advanced by throwing out new threads and curls. A given thread or curl, one moment at the edge, would in a few minutes be well in the band. Quite opposite events were taking place in the rear of each band. The cloud was there obviously melting away. Indeed, to sum the matter up, the cloud bands flowed past their details just as the waves on the sea flow past the floating foam. Evidently we have in Plate 15 the result of a plane of commencing cirrus formation broken into a series of troughs and waves by an undulatory movement of the air. But, as we have already said when speaking of cirrus inconstans, the condition in which trifling up-and-down movements can determine whether condensation shall, or shall not, take place seldom lasts long. It is usually only a stage in a continuous change, and in this particular instance the banded structure was soon replaced by a fairly continuous sheet of typical cirro-stratus.

Plate 15.

CIRRO-STRATUS.

The next plate, No. 16, shows a similar process. In the upper part we have cirrus inconstans forming in patches out of a deep clear-blue sky. Its hazy fibres grow closer and closer, betraying a slight tendency to gather in narrow ripple-like bands, but the structure is soon lost in the uniform white sheet of interlacing fibres, which differ from common cirrus in little else than their number and close-ness. Nevertheless, the stratiform arrangement is quite obvious enough to warrant the use of the term "cirro-stratus."

Plate 16.

CIRRO-STRATUS. (CIRRO-STRATUS COMMUNIS.)

The change of cirro-nebula into cirro-stratus is shown in Plate 4, to which reference has already been made. The structures are remarkably delicate, showing in the middle a distinct irregular mottling; and rather further towards the top right-hand corner a ripple structure appears, and in the top left-hand corner the sheet is denser and whiter. The altitude of this cloud was evidently great, and actual measurement showed it to be 7·6 miles. It did not last long, and after its change into broken patches of denser cirro-stratus, still higher clouds were revealed through the gaps.

Cirro-stratus often forms almost simultaneously at more than one level, and when that happens the full stratiform appearance is generally reached first by the lower layer. In Plate 17 we have two layers. The fluffy bits of cirrus nebulosus, in the lower part of the picture, are really the higher clouds. Below them, probably by many thousands of feet, floats the denser cloud shown in the upper part of the picture. This is an interesting link between the fibrous and the granular forms of cirrus, and is probably best described as spotted cirro-stratus, or cirro-stratus maculosus. It is a form very frequently met with, but seldom showing any persistence. It is indicative of condensation in a calm atmosphere, and not unfrequently marks either the small irregularities of pressure which form the conditions for thunder-storms, or the beginning of the break up of an anticyclone.

Plate 17.

FLOCCULENT CIRRO-STRATUS. (CIRRO-STRATUS CUMULO-SUS.)

A coarser texture and greater density are shown in Plate 18, where we have cirro-stratus in the lower part of the picture, and cirro-cumulus in the upper. The altitude of this cloud was only about 4000 metres, one of the least values recorded at Exeter for cirro-stratus of any kind. The intimate admixture of the fibrous and granular forms is very clearly shown.

Plate 18.

CIRRO-STRATUS AND CIRRO-CUMULUS.

This close relation is equally obvious in Plate 19, where the cloudlets are arranged in loosely marshalled rows, dimly resembling the banded structure of Plate 15. But in this case the direction of movement was with the long lines, and the propagation of cloud production followed the same course. Some of the little cloudlets have an opacity, and therefore brilliancy, quite unusual for cirro-cumulus, but their intimate association evidently in the same plane with undoubted cirrus shows that they must fall under that general description. It is a cloud indicative of unsettled weather, and the exceptional brilliancy is doubtless due to an unusual quantity of vapour at the cloud plane, which must mean that the change from the dry stratum above to the damp one below must be much more sudden than is ordinarily the case. Clouds of this kind might well be called cirro-stratus cumulosus.

Plate 19.

CIRRO-CUMULUS.

We now come to two companion pictures, Plates 20 and 21, which were taken within half a minute of each other. In the first the camera was directed towards the west, and in the second it was facing the north-west. The sun was nearing the horizon, and was only just outside the field of view in each case, so that the two photographs form a panorama of the western sky. A solar halo had disappeared about half an hour previously, and the cirro-nebula had changed into the remarkable forms of cloud depicted. Plate 21 shows cirrus ripples in the upper part, and cirro-cumulus in soft, ill-defined balls in the lower part; but they were at the same level, and are only different parts of the same cloud plane. In Plate 22 we see similar ball-like cloudlets ranged in long lines which run almost at right angles to the ripples of the companion picture. Clouds like these are rare. They are almost unknown during the early part of the day, and, so far as the writer's experience goes, they are only to be found in the afternoon towards sunset. Some of our most gorgeous sunset skies are due to them; for their altitude is considerable, and they do not light up with the sunset colours until the lower clouds have become dark shadows against the glowing background. The hottest months of the year, the still air and great evaporation which are the contributing causes of thunderstorms, are also the conditions under which such skies may be seen. Indeed, while these photographs were being taken, heavy thunderstorms were in progress within less than a hundred miles. Cirro-cumulus nebulosus, or hazy cirro-cumulus, describes the form correctly.

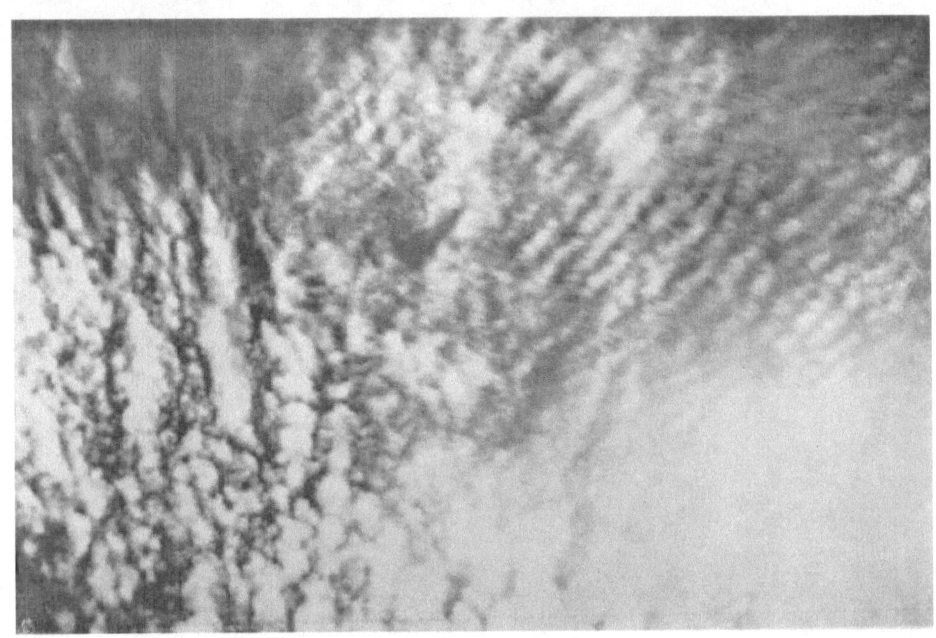

Plate 20.

HAZY CIRRO-CUMULUS. (CIRRO-CUMULUS NEBULOSUS.)

Plate 21.

HAZY CIRRO-CUMULUS. (CIRRO-CUMULUS NEBULOSUS.)

The next plate, No. 22, gives a view of an evening sky about half an hour after sunset. The lower clouds, cirro-cumulus and cirro-stratus, of a deep purple brown, standing out dark against a gold-coloured sheet of higher cirro-stratus, which comes out white in the photograph, while the purple-tinted sky comes dark. We have here three distinct layers, all cirrus. First, the hazy cirro-cumulus, forming two bars across the lower part of the picture; then long bands of cirrus or cirro-stratus, best seen in the bottom right-hand corner; and, far above both, the cirro-stratus which was reflecting the yellow sunlight. Such a sky might be an indication of thunder conditions, or it might be due to an unusual quantity of vapour in the atmosphere produced by some other cause. The actual conditions were the gentle flow over England of vapour-laden air from the western ocean, heralding the change from a long spell of fine hot weather, due to a July anticyclone, to a month of heavy rains and western gales, accompanying the passage of a long procession of cyclones along our western shores.

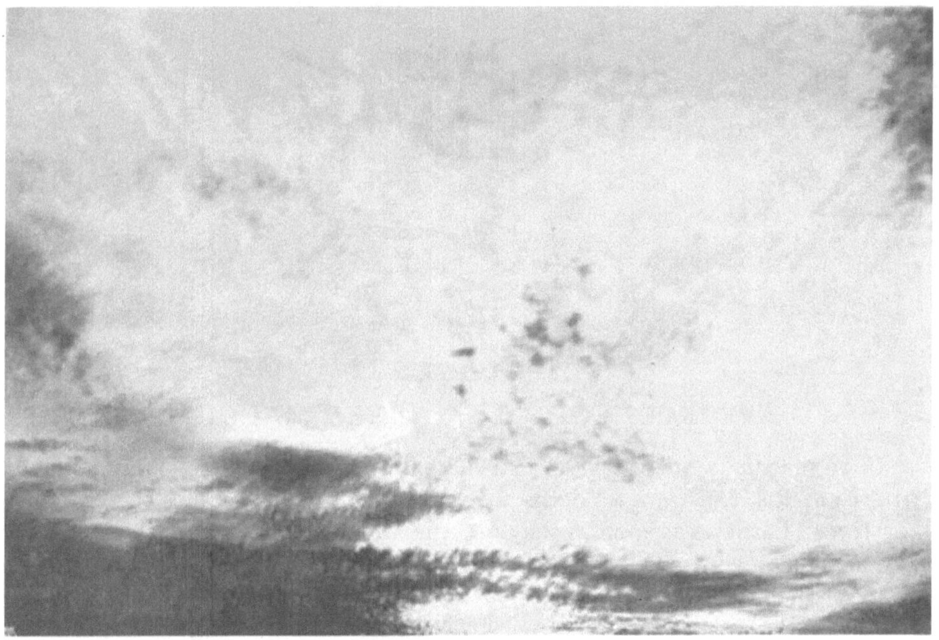

Plate 22.

A SUNSET SKY.

Again, a marked contrast is shown in Plate 23. Here we have the highest and thinnest form of cirro-cumulus, the one named cirro-macula by Mr. Ley. It is rarely, if ever, seen before eleven o'clock in the morning, and is far commoner in the afternoon. The example shown was photographed at sunset at the close of a day which had been almost cloudless. Cirro-macula forms here and there in a clear sky. A hazy, whitish patch appears, which at first shows no definite structure, but looks almost like a little bit of cirro-nebula. This suddenly splits up by clear blue lanes running through it, and cutting the patch up into irregular segments, which quickly round themselves off into minute bits usually whiter on their edges and semi-transparent in the centre. The process can be strikingly imitated by scattering on water some fine powder which will float. If left without disturbance, the particles draw together into numerous small groups, leaving lanes of clear water between them.

Plate 23.

SPECKLE CLOUD (LEY). (CIRRO-MACULA.)

Cirro-macula frequently gives rise to the fibrous form of cirrus we have called cirrus caudatus. The granules of the cirro-macula grow denser, and begin to drop their frozen particles as soon as they become large enough. Indeed, a cloudlet of cirro-macula may sometimes be seen to turn bodily into a fine line of falling crystals, which will be a curving line of cirrus. On the other hand, it will sometimes remain visible for an hour or more without any trace of descending streaks or floating fibres. Pure cirro-macula such as Plate 23 is not often seen; it is far more frequently mixed with more solid-looking cloudlets and descending fibres, such as are shown in Plate 24, which gives the same point of view as 23, but a quarter of an hour later, and photographed with a longer focus lens. These two photographs, together with 20 and 21, give excellent examples of the use of the black mirror. In none of the four could the naked eye detect all of the cloud structures. The whole sky was a blaze of dazzling light, but by adjusting exposure and development the details are fully brought out without the least difficulty.

Plate 24.

CIRRUS CAUDATUS AND CIRRO-MACULA.

Cirro-stratus, we see from the examples which have been considered, hardly deserves to be treated as a distinct genus of cloud. Its formation is identical with that of many species of cirrus, or in some cases with that of the speckle cloud, cirro-macula, or even the coarser kinds of cirro-cumulus. The different varieties which it shows are best rendered by reference to the specific names of the detached forms which have similar features.

Cirro-cumulus, on the other hand, does present clearly marked varieties. Cirro-macula is so distinct that it might well be given the name awarded to it by Mr. Ley, while the term "cirro-cumulus" is reserved for the coarser and rounder forms. The hazy, ripple-like structures of Plate 4 and Plate 20 should also have some distinctive appellation, as will be suggested later on when dealing with wave clouds as a whole.

It is difficult to find any short way of expressing the various ideas which should be summed up in the name of a cloud. There seems no alternative to the use of additional words, unless it be to follow the example of chemists, and compound appalling names similar to those which terrify the uninitiated who think they would like to read something about, let us say, the coal-tar dyes.

If a cloud belongs to the order cirrus, is in a level sheet, and that sheet is composed of interlacing or curling fibres, like those of common cirrus, we can hardly express the facts more briefly than by calling it cirro-stratus communis, or common cirro-stratus. If it consists of cirrus bands fused together, but still showing the banded structure, it is cirro-stratus vittatus. Again, if it is finely speckled, like cirro-macula, it may be described as cirro-stratus maculosus, and if the structure is coarser it may be called cirro-stratus cumulosus.

As a general average, cirro-stratus lies somewhat lower in the atmosphere than the detached forms, probably because the conditions which give rise to the latter reach to greater altitudes in patches than it is possible for them to reach in a continuous manner. Vapour becomes rarer with increased height and with diminished temperature, so that it must, on the whole, be less frequently present in cloud-producing quantity as the height increases. At great altitudes it will be seldom that the quantity is great enough to produce a stratiform cloud, though it may well be enough for cirro-macula, or the detached forms of cirrus, like cirrus excelsus.

The production of cirro-cumulus and cirro-stratus sometimes spreads across the sky with astonishing speed, and this rapid advance of the edge of the cloud may lead to quite mistaken ideas as to the velocity of the wind at that altitude. In the case of cirro-cumulus, or cirro-macula, it is easy to fix attention on a single cloudlet. If this has the usual ball-like form, it can only be regarded as floating in the air and moving with it. Meanwhile new cloudlets may be forming and growing denser, so that the cloud patch as a whole may be apparently advancing at a much greater rate. Careless observation would then lead to the idea that the cloud was moving much faster than it really is, but if the attention is rigidly fixed on a particular cloudlet the mistake is impossible. If the cloud is a variety of cirro-stratus, it is not always easy, or even possible, to distinguish between the advance of condensation and the movement of the whole, but it can nearly always be done if the cloud shows any definite features upon which attention can be fastened. Sometimes none sufficiently marked can be seen, and when that happens it is still possible in most cases, by watching the edge of the cloud-mass, to see whether new cloud is being added to that edge. The wave-like forms present a special case, which will be dealt with in a later chapter, after the general principles of cloud formation have been discussed in connection with the great clouds of the lower air, whose causes and conditions are far better understood.

CHAPTER IV
ALTO CLOUDS

From cirro-cumulus and cirro-stratus we pass through almost insensible gradations to the denser forms classed together in the alto group. These clouds are fundamentally different, in that they are always composed of liquid particles, though there is no doubt, from their great altitude, that their temperature must often be many degrees below the ordinary freezing-point of water. When this is the case, they are not unfrequently more or less mixed with streaks and filaments exactly like those described under the name of cirrus, which have been explained as due to slowly falling snowflakes. It is not immediately obvious how such apparently contradictory statements can be reconciled. The explanation is that minute droplets of water may be cooled many degrees below freezing-point without changing into ice, and that such super-cooled droplets congeal instantly if a few of them join together to form a larger drop. Practically the same process may be watched any day when there is a sharp frost and dense fog drifting slowly along. The fog-particles are liquid, and produce optical effects in the neighbourhood of any brilliant light, like an arc lamp, absolutely the same as those which would be produced if the temperature were above freezing-point, while there are none of the different phenomena which might be expected if the particles were crystalline ice-dust. As these liquid particles drift along they come in contact with branches of trees and other obstacles, the surface stratum which surrounds them and binds them into spheres is broken, and the drop instantly solidifies. It is to be noted, moreover, that the drop does not freeze as such, but merely adds some more particles to the branching crystals of hoar frost, which grow outwards always towards the direction from which the fog is drifting.

Most liquids, when freed from contact with solid bodies, or when surrounded by a smooth envelope of uniform character, can be cooled below their normal freezing-point without solidification taking place; but the introduction of a particle of the solid, or sometimes of any foreign body, instantly brings about a rapid freezing of the whole. These phenomena of surfusion, as it is called, have long been known, and many of them are very interesting and difficult to understand. Indeed, it is probable that we shall have to add largely to our knowledge of the forces which bind the molecules of a body together to form a solid, and which direct the processes of crystallization before we shall be able to interpret with any certainty a series of facts depending on the attributes of those very forces.

Water is no exception. If finely divided, as by placing it in fine capillary tubes, in the pores of wood, or in the narrow spaces of a wick, it may be cooled several degrees below normal freezing-point. In a cloud, or fog, all the conditions necessary for surfusion to take place are undoubtedly present. The water is pure, the envelope is uniform, the subdivision is exceedingly minute, and the drops are free from most of the mechanical disturbances which bring about the solidification of larger masses in the laboratory.

Thus we see there is nothing at all surprising in the fact that clouds composed of liquid particles may exist at temperatures below the ordinary freezing-point. On the contrary, we should expect that the solidification of the cloud particles would not take place until the temperature was many degrees below freezing, as is certainly the case with clouds of the cirrus order. At temperatures between this unknown, but low value, and the normal freezing-point, the clouds will be composed of liquid; but when the particles join together, snowflakes will result instead of raindrops; and this will be just as true of alto clouds as it is of the great vaporous mountains of the lower regions of the air which bring falls of snow. The streaks often mixed with alto-cumulus are cirrus threads, and are, no doubt, of exactly the same nature as the tails of cirrus caudatus, or even the fibres of cirro-filum.

The simplest alto cloud is alto-stratus. When this is complete, so as to cover the sky, it can be distinguished from cirro-stratus by the absence of fibrous structure, and by the facts that it never produces any halo or fragment of a halo, but instead surrounds the sun or moon with a white blur, or, if it is thin enough, with a close ring of coloured light much nearer than a halo, and with the colours in the inverse order—that is, with the red furthest from the centre. Some of these so-called coronæ are very beautiful when seen in the black mirror, and some of those formed around a full moon show quite brilliant tints to the unaided eye. Of course, these meteorological coronæ have no relation whatever to the true solar corona; they are simply formed by the passage of the rays of light through the veil of small particles, and may be easily imitated. Take a piece of glass such as a lantern-cover glass, breathe on it, and hold it close before the eye while looking at some small source of light. If the dew deposit is thin, bright colours are shown in a luminous ring surrounding the light, and the thinner the deposit of dew the larger the ring will be. Breathe heavily so as to give a thick deposit, and the light will be seen to be the centre of a patch of white brightness without any colour.

The phenomena are due to what is known as diffraction, and if the other conditions are unchanged the diameter of the ring is inversely proportional to the size of the particles. Purity of colour in these rings is an indication of uniformity in the size of the particles. When the moon is shining through a sheet of alto-stratus, which thins off to one edge, very beautiful effects may often be noticed, and the change from the colourless blur, when a thicker part of the cloud is interposed, to the brilliant colours of the corona formed by the thinner edges is very striking. Similar phenomena are shown almost equally well by any of the alto clouds, but cirrus thin enough to produce a coloured corona will generally produce a halo.

Alto-cumulus of the kind most nearly allied to cirro-cumulus is shown in Plate 25. The upper part of the picture shows ragged, irregular patches, with slight indications of fibrous streaks. The lower portion shows rounder, ball-like cloudlets, a few of the larger of which have distinct shadows on the side away from the sun.

This plate gives alto-cumulus in a partly formed condition, but it is not a mere passage form. Sometimes exactly such a cloud will float overhead for hours, showing very little movement and only slow changes of detail. It is therefore a distinct variety, and may be called alto-cumulus informis.

Plate 25.

ALTO-CUMULUS INFORMIS.

A less definite form is shown in Plate 26. It may also be regarded as only partly formed, but its construction is quite different, every part being misty and ill defined. It is a common cloud, especially in sultry summer weather with still air. Under those circumstances, after a hazy morning, it may be seen slowly forming during the afternoon, growing in density as the hours go by, until it reaches a maximum about five or six o'clock, after which it melts away, or settles down into small patches of high stratus. Most frequent in summer, it is by no means rare in autumn and winter, but still air is essential. From its hazy appearance it may be called alto-cumulus nebulosus.

Plate 26.

HAZY ALTO-CUMULUS. (ALTO-CUMULUS NEBULOSUS.)

Fixity of detail and slow movement characterize both the foregoing forms, and in that respect our next picture (Plate 27) shows a cloud which is a great contrast. Its detached cloudlets are rather flatter and thinner, and though the cloud as a whole will often persist for hours, it is undergoing continual change, and is formed when the air is far from still. Cloudlets form and gather into stratiform patches, which soon break up again and disappear; and the process goes on here and there, sometimes accompanied by fairly rapid movement of the patches as a whole. This cloud may be described as alto-cumulus stratiformis.

Plate 27.

FLAT ALTO-CUMULUS. (ALTO-CUMULUS STRATIFORMIS.)

We now come, in Plate 28, to a cloud of singular beauty. It forms rapidly in a clear sky, its first traces bearing a striking resemblance to cirro-macula, but the floccules, instead of remaining semi-transparent or dropping cirrus threads, rapidly become opaque balls of cloud which lengthen upwards. This upward tendency causes the formed cloudlets to have their longer axes vertical, which is very characteristic. It might be named alto-cumulus castellatus, or high-turreted cloud. Mr. Ley named it stratus castellatus, or turret-cloud, but it certainly belongs to the cumulus section of the alto group. Thunder weather is the invariable condition for its production. If it is seen, at least in England, thunderstorms are certain to be recorded not very far away. When this particular photograph was being taken in South Devon, very destructive storms were recorded in Brittany and in the English Midlands, and the anvil-shaped tops of unmistakable thunder-clouds were visible above the horizon while the exposure was being made.

Plate 28.

HIGH TURRETED CLOUD. (ALTO-CUMULUS CASTELLATUS.)

Another form of almost equal beauty is shown in Plate 29. The rounded balls make their appearance as semi-transparent spots upon the sky, and in their general characters might easily be mistaken for cirro-macula. But a few minutes will be enough to decide the question. The little spots rapidly grow denser, frequently becoming ragged at the edges; they never drop down the slender filaments which usually descend from cirro-macula, and their edges are never denser than their central parts, which, it will be remembered, was a frequent feature of the true speckle cloud. The cloudlets are obviously rounded balls arranged in patches, which may turn gradually into alto-stratus by their fusion, or, after an existence of minutes or hours, the whole may disappear by a disintegration of each ball, by its breaking up into a ragged mass and melting away. The altitude at which this cloud forms is between 5000 and 9000 metres, according to measurements made by the writer, the actual specimen figured being about 7000. It is almost as characteristic of thunder weather as the last, but whereas Plate 28 shows a variety which is most often seen before 3 p.m., since it only occurs while the cloud planes are rapidly rising, the one before us may be formed at almost any time of day, but most frequently occurs in the afternoon. An imperfect form of it is frequently met with about sunset, in which the rounded balls are not usually so well defined as when the sun is high above the horizon. Alto-cumulus glomeratus would be a suitably descriptive name.

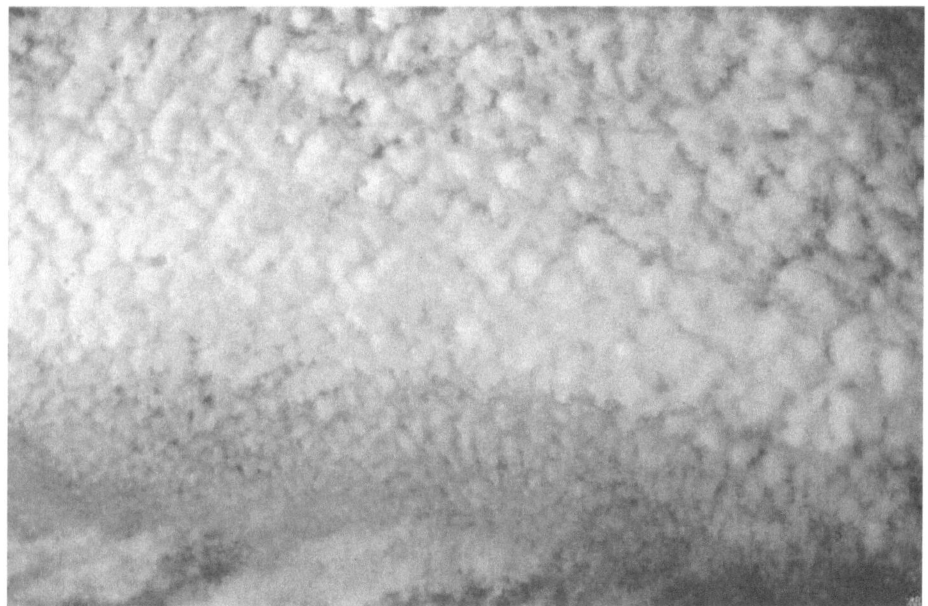

Plate 29.

HIGH BALL CUMULUS. (ALTO-CUMULUS GLOMERATUS.)

If it were possible to take a good typical example of the variety just described and roll it flat, so that each cloudlet should be reduced to a lenticular shape, we get a type which seems seldom to appear during the heat of the day, and to be most frequent about sundown. It consists, as shown in Plate 30, of distinct cloudlets, with considerable spaces between them, and gives the impression of a discontinuous level sheet. But the component cloudlets are much too definite, and preserve their individuality far too well to suggest any idea of a broken stratus; the spotted structure is the predominant feature, while the stratiform arrangement is almost equally plain. Alto-stratus maculosus would be a suitable term. It is not so high a cloud as the glomeratus type, the one shown being at an altitude of about 5600 metres. The plate shows the position of the setting sun, which is partly hidden behind some dark patches of broken alto-stratus (fracto-alto-stratus), the hazy form and boundaries of which form an effective contrast to the shining cloudlets 2000 metres or so above them. Many of our most beautiful sunsets are due to this form of cloud, particularly in the late autumn. It is a cloud of calm weather, and often floats apparently motionless, and undergoing little change, like flakes of glowing fire against the background of a fading sky long after the sun has disappeared. It is not indicative of thunder conditions, and it may occur on the margins of an anticyclone.

Plate 30.

MACKEREL SKY. (ALTO-STRATUS MACULOSUS.)

A lower and coarser form of the spotted alto-stratus is shown in Plate 31, where it is seen through the gaps in a thin sheet of broken stratus. In this case also the sun was getting low in the sky, being hidden by the denser bit of stratus in the bottom left-hand corner.

Plate 31.

MACKEREL SKY. (ALTO-STRATUS MACULOSUS.)

Alto-stratus does not often, if ever, grow from the fusion of the cloudlets of the maculosus type. But it does come from alto-cumulus glomeratus, and also from a form shown in Plate 32. Here we have alto-stratus in process of growth. Small irregular lumps of cloud forming on the right-hand side of the picture grow larger and more irregular, begin to fuse together towards the centre, and on the left-hand side the fusion is almost complete. Still, although the sky is covered with cloud, the lumpy form is plainly visible. The term "alto-strato-cumulus" is suitable, as it differs from the more frequent and much lower cloud, which will be described further on as strato-cumulus, in little else than altitude and general massiveness of texture. This high strato-cumulus is common enough, too common, indeed, in England, as it produces many a dull grey sky both in summer and in winter. In the latter season it is not unfrequent with the cold east winds of February and March. It is probably the lowest of the alto clouds; the lowest measurement made by the writer being 1828 metres at Exeter, but lower altitudes seem to have been recorded elsewhere.

Plate 32.

ALTO-STRATO-CUMULUS.

Alto-cumulus castellatus, which is breaking up and disappearing, is shown in Plate 33. It was photographed with a long-focus lens, so that the scale of representation is about eight times as great as that of Plate 27. This view was taken at Exeter while a thunderstorm was in progress at Bristol.

Plate 33.

SUNSET. (ALTO-CUMULUS CASTELLATUS FRACTUS.)

CHAPTER V

LOWER CLOUDS

THE clouds of the lower portions of the atmosphere are formed in regions where water vapour is abundant, and frequently in easy reach of the strong ascending and descending air currents produced by the varying temperatures and irregular surface of the ground. It is sufficient to recite these conditions to show that these lower clouds will be denser, larger, coarser in texture, and characterized by greater definiteness of form than those we have, so far, considered.

In the International system they are classified thus—

Group C. Lower clouds.

    (*a*) Strato-cumulus.
    (*b*) Nimbus.

Group D. Clouds of diurnal ascending currents.

    Cumulus and cumulo-nimbus.

Group E. High fogs.

    Stratus.

This is certainly the least satisfactory part of the whole scheme, and it is not at all easy to see upon what grounds it was adopted by the International Committee. Group D—cumulus and cumulo-nimbus—do show important differences from the other groups, though it is often difficult to say whether the sky should be described as covered with strato-cumulus or as covered with numerous small cumulus. It is the separation of stratus—placing it in a group by itself, and making that the lowest—which is the worst point. As a matter of fact, stratus may exist at any altitude from sea-level up to such heights that we should not hesitate to call it alto-stratus. Indeed, there is no essential character of alto-stratus which distinguishes it from some of the lower forms. Whatever its altitude, its thickness and the size of its particles may vary in a precisely similar manner. We may have the particles exceedingly small, when the fog will be dry, and such a stratus may be so thin as hardly to dim the sun; or it may be so thick as to completely hide it. On the other hand, the fog may consist of particles easily visible to the naked eye, forming the so-called Scotch mist, or the "dry" fog of Dartmoor, which will wet things as rapidly and more thoroughly than a smart shower. When such a fog accumulates to a sufficient depth, the particles in their fall pick up others, and the result is a distinct fine rain. This may occur not only near the ground, but at almost any level below that at which the cloud would pass into the region of cirrus.

Plate 34 shows three layers of stratus, in each case much broken up. The highest layer is a good example of alto-stratus maculosus. Lower down, by half a mile or more, come parts of a grey sheet considerably denser and thicker. It is a matter of taste whether this should be called high stratus or low alto-stratus. There is no test by which the one can be distinguished from the other. Lower again come the detached darker clouds, which are fragments of a sheet of stratus which is breaking up and disappearing. The photograph was taken in the afternoon, after a wet morning, and all three layers were probably relics of the great rain-cloud system, or nimbus, which produced the rainfall.

Plate 34.

THREE LAYERS OF STRATIFORM CLOUD AFTER RAIN.

We have referred to the production of fine rain from a thick fog. If now such a thick layer of coarse-grained fog—if we may use such a phrase—is suspended overhead at a moderate altitude, the result is a drizzling rain underneath, and the cloud at once becomes a nimbus. When Luke Howard adopted the term "nimbus," he proposed to employ it, apparently, for a vast mass of cloud such as that which forms the rainy region of a cyclone; a huge pile of clouds containing representatives of all his other types in some unknown but close relationship. It was, in fact, a comprehensive term, and as such there was a good use for it. At present it is applied to any cloud from which rain is falling, except when the cloud can be identified as a variety of cumulus which is called cumulo-nimbus. But we have already said that a stratus may be a rain-cloud, and so may other varieties. Moreover, whenever a nimbus breaks sufficiently for us to be able to see its upper surface, we invariably find that, if it were viewed from above, we should, without a moment's hesitation, place it in one of the other groups. It is only when we are underneath it we can see its rain-producing character, and give it the orthodox name. The real fact is that nimbus should be an adjective, meaning rain-producing, and not a substantive.

However, it has its allotted place in the International system, and it is better to adhere as far as possible to a defective but widely recognized system until it can be authoritatively amended, rather than to make an individual attempt to ignore it. The facts are sufficiently obvious, and the days of nimbus as a type are numbered. The two plates, Nos. 35 and 36, are fair typical representations of the clouds usually known as nimbus; but they are both of them only the under-surfaces of other clouds, Plate 36 showing the under-surfaces of a group of heavy cumulo-nimbus all joined together so as to cover the sky, while Plate 35 shows a mass of dense strato-cumulus. The rain-cloud is always a form of either stratus or cumulus, or a combination of the two, sometimes in further combination with clouds of the alto class, or even extending upwards to cirro-stratus and cirro-nebula. Where it consists of a single layer, that layer differs from its rainless representative only in greater thickness from base to summit, or in greater density; and when there are several distinct layers of cloud, so that the lowest is shaded by the higher, rain may fall, even though they differ in no visible way from clouds which would be rainless if alone. Plate 33 is an example.

Plate 35.

RAIN-CLOUD. (NIMBUS.)

Plate 36.

RAIN-CLOUD. (NIMBUS.)

Nimbus, indeed, is not a type-form, but is merely a typical condition, and when used as a substantive is only a convenient way of expressing our ignorance as to the real form of the cloud we so describe.

The altitude of the base of a rain-cloud may vary considerably. It may be anything from sea-level up to heights which vary with the geographical conditions and with the conditions of temperature and pressure, but probably in this country never greater than 7000 or 8000 metres.

Rain, or snow, often falls from clouds at greater altitudes than these, but unless in its descent it passes through other lower clouds, the drops, as a rule, will dry up and disappear. The author has often seen quite heavy rain descending from a cloud, and disappearing completely within a thousand feet or so of the cloud-base. On rarer occasions a still more remarkable thing may be seen—namely, a shower falling from an upper cloud into a lower, and none between this lower cloud and the ground. This curious phenomenon can only be explained by supposing that the convection currents which make the lower cloud are strong enough to support the small raindrops.

Pure stratus is a level sheet of cloud with little variation of thickness, not ascending every here and there into rounded lumps. Its most typical form covers the whole sky with a uniform grey pall, which may or may not completely hide the sun. Such a cloud does not lend itself to pictorial representation. A frequent form, in which the sheet is more or less broken, is shown in Plate 37. This is a variety which is frequent in the summer mornings, and generally breaks up and clears away before eleven o'clock. If, however, it appears in autumn and winter with layers of alto cloud above, it may grow denser, and turn into a stratiform nimbus, or it may go on drifting overhead for several days without sign of change.

Plate 37.

STRATUS COMMUNIS.

Break up such a sheet of cloud by numerous meandering cracks, and round off the detached pieces so as to give them a more or less rounded or pyramidal section, and the cloud becomes strato-cumulus, typical representations of which are shown in Plates 38 and 39, which depict different parts of the same sky. In Plate 38 the camera was pointed due west, and in Plate 39 it was turned round to the north-west, so that the two views do not quite meet.

Plate 38.
STRATO-CUMULUS.

Plate 39.
STRATO-CUMULUS.

Plate 40 is a different variety. It is stratiform, each component cloudlet being rather ragged at the edges. In some ways it resembles cirro-macula and the speckled varieties of alto cloud, but it is coarser in texture and obviously at no great altitude. The International system would call it strato-cumulus, but Mr. Ley gives a representation from another negative taken at the same time as the type of what he calls stratus maculosus, a name which seems far more suitable, since the cloud bears a much closer relation to stratus than to cumulus. In the particular instance figured, the broken structure did not last long; the spaces gradually closed in, and a complete stratus was the result.

Plate 40.

STRATUS MACULOSUS.

Strato-cumulus often lasts for hours, with little or no perceptible change, but stratus maculosus rarely persists for more than half an hour. The first is a cloud of fairly stable conditions, the latter is dependent for its existence upon the near approach to critical conditions at one particular level, and, as we have said in other cases, such a critical state is almost always soon passed, with the result that the cloud either masses into a denser form, or else breaks up and disappears. If the up and down currents are strong enough to persist, the result will be strato-cumulus and not stratus maculosus.

A kind of stratus which is frequently seen in the daytime is shown in Plate 41. This is literally a lifted fog, having been formed about midday, after ground fog in the early morning. It would be called common stratus, or stratus communis. When it appears it is a fairly persistent form, sometimes breaking up or swelling up into strato-cumulus, but more often splitting into long rolls of cloud, with margins like those of cumulus. This phenomenon is shown in Plate 42, which was taken in December at 11 a.m., on a day which opened with a thick ground fog. A precisely similar cloud is frequent in the early hours of a summer morning, as a stage in the dispersal of a radiation ground fog. The fog first lifts from the ground, until it reaches a height of a few hundred metres, when it splits into the long rolls whose axes are at right angles to the direction of drift. The consequence is very strange if you stand on a hilltop close under the drifting mass, and look towards the horizon in the direction of drift. The changing shadows give the impression that the clouds are actually rolling along, though of course no such thing is really taking place. As time goes on the rolls grow larger and the interspaces wider; then transverse fissures appear, and gradually the rolls break up into small detached cumulus. Cumulus radius, from the Latin for a rolling-pin, might be a suitably descriptive name, but it should not be forgotten that it is only an intermediate link between stratus and cumulus, and, indeed, is more nearly related to the former, since it is never produced except on the break up of stratus, while it may dry up and disappear without reaching the cumulus stage at all. Stratus radius would therefore be a better name.

Plate 41.

COMMON STRATUS. (STRATUS COMMUNIS.)

Plate 42.

ROLLER CLOUD. (STRATUS RADIUS.)

Cumulus is closely related to another form of stratus, which Mr. Ley has named stratus lenticularis, but this appears to be so frequently the last stage in a disappearing cumulus that its history will come better later on. It is mentioned here as it is, after all, one of the commonest of all forms of stratus, the form which appears at, or after, sunset, and is one of the few clouds which have an English popular name—Fall cloud. Plate 47 gives a representation of it, standing out dark against an evening sky, with a sheet of alto-stratus far above it in the upper part of the photograph.

To sum up, then, we have among the lower clouds of more or less stratiform pattern—

        Stratus communis, or Common stratus.
        Stratus lenticularis, the Fall cloud.
        Stratus radius, or Roll cloud
        Stratus maculosus, or Mottled stratus.
        Strato-cumulus, or Sheet cumulus.

The last leads naturally to the consideration of cumulus and cumulo-nimbus, while the term "nimbus" does not belong to any one type-form, but sometimes to one, sometimes to another, and generally to a mixture of two or more.

A good many years ago the writer made a series of measurements of the thickness of detached clouds of the stratus and cumulus types, such as those which may produce a shower. The conclusions reached in consequence of those determinations have since been amply confirmed by subsequent observations. In winter no rain will fall from a cloud unless it reaches a minimum thickness of at least 100 metres, while in summer it must have rather greater thickness. There is one exception, and that is in winter, when the temperature is so low that the drop starts on its downward journey as a flake of snow. When this is the case, rain may fall from a layer of thin lifted fog, not quite thick enough to hide the blue colour of the sky. But under ordinary conditions of temperature, if the cloud has a thickness less than 2000 feet, or 616 metres, rain is unlikely, but if it does come, the drops will be small and the fall of rain quite trifling.

Above this thickness the heaviness of the rain and size of the drops increases, so that if the distance from base to summit be between 2000 and 3000 feet, or 600 to 1000 metres, the fall will be gentle. A thickness of 4000 to 6000 feet, or 1200 to 1800 metres, gives large drops and a fairly heavy shower, while, in summer time at least, cold heavy rain and hail come from clouds measuring 6000 to 10,000 feet, or in round numbers 1800 to 3000 metres or more. In winter the necessary dimensions seem to be less, but the rule still holds equally good, that the rain-cloud does not necessarily differ in any way from the rainless one, except in thickness, and that when the requisite thickness is present rain is not always the result.

CHAPTER VI

CUMULUS

UNDER the general term cumulus there are grouped the most common, the best known, and the grandest forms of cloud. Indeed, beautiful as the cirrus and alto clouds may be, there is a solid grandeur about the greater forms of cumulus which gives them a beauty of their own quite comparable with the charm afforded by the delicate tracery of their more lofty rivals.

Cumulus can be divided into several types, which are best considered in the order of growth. They are all formed in the lower part of the atmosphere, their under-surfaces varying in altitude from about 600 metres, or even less, up to 3000 metres, or slightly more. The writer's own measurements vary from a minimum of 584 metres to a maximum of 2286 metres, with an average of a little more than 1000 metres.

They are described in the International system as "clouds in a rising current," and there is no doubt the description is correct. Each cumulus must be looked upon as simply the visible top of an ascending pillar of damp air. The vapour which makes its appearance in the cloud is present in the transparent air beneath, and the base of the cloud is simply the level at which that vapour begins to condense into visible liquid particles. Since cumulus clouds are caused by ascending currents, these currents must be brought about either by the general disturbance of the air due to a cyclonic movement, or by the local irregularities of temperature on the ground produced by the sun's heat. As a matter of fact, we do get cumulus produced in great abundance in the rear of every cyclone, and we get them also under the conditions of still air and hot sun, which specially favour evaporation and the development of differences of temperature. The cyclone cumulus may come at any hour of the day or night, though comparatively rare between midnight and the morning. Heat cumulus is generally formed during the afternoon, and it is only under relatively uncommon conditions that it persists during the night. If the cloud has not grown to very great size it usually begins to break up and disappear about sunset, but if it has grown to the enormous dimensions of a summer thunder-cloud it may go on growing, piling mass on to mass, until it generates a thunderstorm, even in the hours of early morning.

In the case of some of the higher kinds of cloud, we are not able to give any certain account of the mechanics of their production from a study of those clouds themselves. We have already referred incidentally to some of the speculations as to their origin and some of the facts definitely known, but considerable light can be thrown on the genesis of all the varieties of cirro-cumulus and alto-cumulus by a careful study of their larger and more accessible representatives of lower regions.

The cyclone cumulus does not differ in any essential from the clouds of calm weather. The only difference is that the uprising currents are perhaps partly eddies, and the rate of fall of temperature with ascent is often more rapid.

Given any mass of air at a particular temperature, it can take up and hold in the form of invisible vapour a fixed quantity of water, and no more. When it holds the maximum possible it is said to be saturated. If it is nearly saturated it would be called damp; if far from saturated, dry. Now, the warmer the air the larger the quantity of vapour necessary to saturate it, so that if a quantity is saturated at a high temperature, and is then cooled, it will no longer be able to retain all its moisture in the invisible form, but the surplus quantity will make its appearance as liquid particles, that is to say, as mist or cloud.

Similarly, if a quantity of air is not fully saturated at its particular temperature, and is then cooled, it will approach nearer and nearer to saturation, and if the process is continued long enough the result will be cloud formation.

All clouds, without exception, are produced by exactly such cooling of air containing water vapour, first to the temperature at which the quantity it contains is the maximum possible, and then beyond that point. Now, if we start with very warm air, and cool it 1 degree, we decrease its vapour-holding power, and the decrease per degree grows less and less as the temperature falls. Suppose, for instance, we have air saturated at 61 degrees and cool it to 60 degrees, the quantity of vapour condensed will be equal to the difference of holding power. Suppose, again, we have air saturated at 31 degrees and we cool it to 30 degrees, the quantity of vapour condensed will again be equal to the difference of holding power; but this quantity will be very greatly less than in the former case. Cooling air saturated at 61 degrees to 60 degrees might produce a dense cloud; but applying a similar reduction of 1 degree to air saturated at 31 degrees, if we take the same volume of air, will only produce a very much thinner result. Here we see one good reason why the highest clouds are the thinnest and the alto clouds of intermediate density.

The necessary cooling may be brought about in several ways. Firstly, the air is capable of radiating its heat into space, and therefore of cooling. But we know little of the laws which govern atmospheric radiation, and presumably, if cloud could be produced by such means, it ought to make its appearance most frequently in the small hours of the morning before sunrise. We are, however, unaware of any variety of cloud which answers those conditions, unless it be the ground fogs which so often form during the night; and these, we know, are certainly due to the chilling of the air by contact with the ground, which has been cooled by radiating away its heat. On the contrary, it is well known to astronomers that the sky is, on the whole, clearer and freer from clouds after midnight than in the earlier hours of the night—a circumstance which is particularly unfortunate for the amateur star-gazer, who has to be up and about at the same time as the rest of the working world. Cooling by radiation we may then dismiss as a cause of cloud formation of no great efficacy, and certainly one which has little to do with the production of cumulus.

Cooling by contact with a cold body is another and more potent cause. We often see it in a mountain district, where a frost-bound peak stands facing the wind with glittering snow-slopes on which the sun is shining, while a long tongue of cloud hangs like a banner on its leeward side. In such a case it is easy to understand how the air sweeping by the icy mass is chilled below its saturation point; but as it passes on, the chilled portions become mixed with the rest, and the cloud evaporates again. It is not quite so easy to see how far this cause is responsible for the clouds which are formed when the warm damp air of the ocean drifts over a comparatively cold land. It is probable that the contact chilling is in this case only part of the explanation, and that other causes co-operate.

The mixing of warm damp air with cold has often been adduced as a cause of clouds. No doubt it might be, and some of the stratiform types may possibly be formed at the junction between a warm damp stratum of air and a cold one, but no example is certainly known. It may also be a contributing cause in producing the sharply defined upper surfaces of some cumulus or strato-cumulus clouds, but these are in the main most certainly due to the chief cause of cloud production—namely, what is known as dynamic cooling.

If a quantity of air exists under a certain pressure and at a certain temperature, on reducing the pressure it will expand, and in the act of expanding it will become cooler. This may easily be illustrated with an air-pump. Let a damp sponge or a piece of wet blotting-paper stand under a glass receiver over an air-pump until the air has become damp. If the apparatus is in a darkened room, and a powerful beam of light from a lantern is sent through the receiver, the damp air will be seen to be quite clear; but a stroke or two of the pump removes some of the air, the remainder is chilled by its own expansion, and a dense cloud is precipitated. If this cloud be viewed closely, it will be seen to be composed of minute particles, which, on looking towards the light, glow with the colours of a corona. In a few minutes the cloud will disappear, but it can be recalled again and again by successive strokes of the pump, getting thinner and thinner as the air gets more and more rarefied; an illustration of a second reason why the high clouds are thinner than the lower.

Some years ago Mr. John Aitken showed that if the damp air used in this experiment were carefully filtered, so as to remove all foreign particles, no cloud was produced, and the introduction of a puff of unfiltered air was attended by immediate condensation. The deduction was that vapour, even below its saturation temperature, cannot produce cloud unless nuclei of some sort are already present, presumably dust particles. Later on it was shown by Mr. Shelford Bidwell and others that gaseous particles, such as those produced by the burning of sulphur, would serve the purpose, and that the brush discharge from an electrified point was in some mysterious way particularly effective. It has recently been shown by Mr. C. T. R. Wilson that causes such as the radiations of radium, or the impact of ultra-violet rays, acting on the air itself, splits up some of its particles into the smaller bodies known as ions, and that these are efficient nuclei. These experiments open up many most interesting questions, but, unless it is to explain the extreme density and darkness of a thunder-cloud, they do not seem to play any important part in determining the forms to be assumed. Nuclei in sufficient abundance are probably always present at any height which can be reached by enough vapour to form a cloud.

Now, if we have a quantity of air, say at sea-level, damp but not saturated, and it is caused to ascend, either because it is warmer and therefore lighter than the surrounding air, or for some other reason, as it moves upwards the pressure upon it will decrease, it will expand, and in the act it will be steadily cooled. This cooling may after a time bring it down to the same temperature as the rest of the air at its particular level. If so, it will no longer be lighter, and the ascent will come to an end. But before this state of affairs is attained it may have reached its saturation point, and cloud production will begin.

It is true that the rarefaction of the air tends to enable it to retain more vapour than it could if it were cooled without change of density. The temperature of the air being fixed, its holding power increases with decrease of pressure. But this increase is much less than the diminution due to cooling, and the result in nature must be similar to what we can see happen under the receiver of the air-pump.

The condensation of water introduces another factor of great importance. It has just been said that the ascending air may be cooled so rapidly as to be reduced to the same temperature as the rest of the air at that level, and if so the ascent will end. Clearly the cessation or persistence of the upward motion depends upon whether the diminution of temperature per 100 metres of ascent is most rapid in the rising column or in the air outside it. As long as the ascending air is warmer than that outside, but at its own level, so long will ascent continue. Now, as long as no condensation was taking place, the rate of cooling would follow a simple law which produces a cooling of 1 degree for about 100 metres of ascent; but as soon as water vapour begins to pass into the liquid form, a large quantity of heat is set free, and the rate of cooling is consequently greatly lessened. Cloud production tends, therefore, to accelerate ascent, and the greater the amount of condensation, the more important will this consideration become; though, on the other hand, when once the cloud is formed, it tends to stop the rising current by shading the air and ground beneath it.

On an ordinary day the rate of decrease of temperature as we ascend is rather less than the value given above, and uprising currents are soon checked. If they do extend far enough to reach cloud production, the clouds will be small, forming the smallest variety of cumulus. This is shown in Plate 43. Small irregular uprising currents have just been able to reach far enough up to have their summits tipped with cloud.

Plate 43.

SMALL CUMULUS. (CUMULUS MINOR.)

After the foregoing explanation, it is easy to see why at a given time the floating cloudlets should have a common base level. This is the height to which the air must attain before reaching its saturation temperature. Each cloudlet marks an uprising current, and the intervals show the position of the counterbalancing descending streams.

A larger variety is shown in Plate 44. In this the level base and generally pyramidal shape is shown, and also the hard, rounded upper surface. The thickness of this cloud was about 500 metres. When clouds like these are visible, they may be the beginning of larger ones, and the only way to judge whether they are likely to develop into rain- or shower-clouds is to watch them. If they are seen to be growing larger, and particularly if detached fragments are developing into clouds, further growth is almost certain, and rain is probable.

Plate 44.

CUMULUS.

If great towering masses are making their appearance with little dark fragments between them, as shown in Plate 45, then smart showers may be confidently expected. The cloud figured was a shower-cloud, and the distance is seen through the veil of falling rain. The height and thickness of this particular cloud were measured just after its photograph had been taken. Its base was 1200 metres above the ground, and its summit was 1500 metres further. Its thickness from summit to base was, therefore, not much short of a mile, and the total contents of the cloud were probably between one and a half and two cubic miles. The upper contour is hard and rounded, as in the smaller cloud of Plate 44, but the whole cloud is much larger.

Plate 45.

LARGE CUMULUS. (CUMULUS MAJOR.)

We have already explained that there seems to be a definite connection between the thickness of such clouds and the amount of precipitation from them. Small cumulus, less than 120 metres thick, rarely produces rain, and nothing like a heavy shower is likely unless the thickness exceeds 400 metres. In winter, especially in hard frost, snow crystals may fall from the smallest cloud, even from little fragments only a few metres thick, but the quantity of water so precipitated will, of course, be small.

As long as the top of the cumulus is rounded and clearly defined, the conditions of aërial equilibrium are stable, and the growth of the cloud has been brought to an end by a stoppage of the ascending current. In Plate 45 the ascent has been hindered both by the mechanical action of the falling raindrops and by the cooling of the lower parts of the ascending column by the descent into it of the cool drops from its colder upper part. This is probably one of the chief reasons why a shower-cloud never maintains its activity as a rain producer for more than a very limited period. As the cloud drifts over the landscape, it seldom maintains its showery character for more than ten or twenty miles, often for much less.

Cumulus, like any of these three, is a cloud of the daytime. It generally begins about ten or eleven o'clock in the morning, grows larger until about four o'clock, and then begins to break up and disappear. After the ascending currents have ceased, the component cloud particles slowly settle down into the warmer air beneath, until the mass has lost its proper pyramidal form, and has become an irregular cloud, such as is shown in Plate 46. This is known as degraded or fracto-cumulus.

Plate 46.

FRACTO-CUMULUS.

One consequence of the arrest of the uprising currents is the formation of lenticular patches of stratus, called by Mr. Ley stratus lenticularis. This is often formed about sunset, and has been named fall cloud, from its appearance at the fall of night. The name is appropriate in another way. The ascending currents having ceased, the cloud particles slowly subside until they dry up in some warmer stratum. The water vapour does not continue its descent, but slowly diffuses in all directions, and if the fall of cloud particles is sufficient, this stratum, which is approximately coincident with the base of the original cumulus, soon becomes saturated, and further particles which fall into it remain visible. This saturated zone will slowly sink lower and lower with the descent of the particles, until it reaches regions in which the temperature is high enough for the whole to be evaporated without reaching saturation point. Evening stratus in calm weather always goes through this sequence of changes. It usually forms at, or soon after, sundown, and begins to break up and disappear as the stars are becoming visible in the darkening sky. Plate 47 shows a specimen of this evening stratus.

Plate 47.

FALL CLOUD. (STRATUS LENTICULARIS.)

A curious feature is sometimes shown on the underside of a thick cloud, which is probably due to the upper part of the ascending column having been carried beyond its position of equilibrium by its own inertia, and then falling back again in the teeth of the still rising lower part. The result is to give the base of the cloud an appearance of a number of rounded masses hanging downwards below the cloud, very suggestive of the idea that the cloud is upside down. Such an event will not often occur, and when it does the conditions are quite wanting in stability, and the consequent features will be very transient. When the base of a cumulus or cumulo-nimbus is so affected, the cloud is known as festooned cumulus, or cumulus mammatus. A precisely similar structure may be seen under strato-cumulus, or even thick stratus. In some countries it seems to be frequently observed, but in England it is so uncommon that the writer has only noted it about a dozen times in twenty years, and on no one of these did it last long enough to allow of its portrait being taken. It is an indication of very disturbed conditions, and is usually followed by heavy rain.

When cumulus clouds are formed in air which is steadily moving as a whole, that is to say, when there is a steady breeze, they have a very decided tendency to follow each other in long lines. It may often be noticed that in a particular place with a certain direction of wind these long processions follow definite tracks in relation to the geographical features. The phenomenon does not seem to have been recorded except in hilly country, but has frequently been observed by the writer. It is not the same thing as the formation of stationary belts of cloud transverse to the wind. These cumulus float along with the movement of the air, and the question to be answered is, why should they follow each other so persistently, and why should the intervening belts of sky be so continuously free from cloud.

If we consider that the warm damp air which supplies them is drawn from the ground, it seems that any cause which tends to direct this warm stratum into definite channels, as it is carried on by the wind, will be a competent cause of the whole phenomenon. This we find in the presence of lofty hills which stand in the way of the warm surface winds, causing them to follow more or less the general trend of the valleys, and so delivering the rising convection currents of cloud-producing air at the same spot.

It is easy to conceive that other causes, such as a difference in temperature or dampness of neighbouring tracts, resulting from whether they are bare or wooded, marshland or sandy plain, might equally suffice; or might, at least, powerfully co-operate with, or counteract, the effect of hill and vale. But in any case it is plain that the geographical conditions to the windward of the place of observation not only may affect the occurrence and distribution of cloud, but if the wind is steady it is difficult to see how they could avoid affecting it.

Another puzzling phenomenon, sometimes presented by cloud and fog, is that our instruments for detecting humidity show that the air within them is not always fully saturated. It seems probable that this is due to such cloud or fog having begun the process of drying up, or that in some way not fully understood the presence of the cloud particles after they have first come into existence may cause the withdrawal of some of the moisture from the intervening damp air. The surface of each minute droplet exerts a pressure on its interior similar to the pressure exerted by the film of a soap-bubble on the air within it, and it is conceivable that some of the uncondensed vapour from outside may diffuse through this enclosing surface film, and be retained there in consequence of the pressure. If this is so, and subsequent investigation can alone decide the matter, it will follow that when once cloud production has begun it will be continued until the air between the cloud particles is reduced so far below its saturation point that the tendency of the drops to evaporate, that is to say, for the imprisoned water to escape through the confining film, balances the retaining pressure.

This consideration, however, is quite incompetent to affect the general explanation of cloud formation which has been given. Its result would be to carry condensation a little further than the exact saturation point, and to retard equally slightly the subsequent evaporation of the cloud particles.

We have spoken of the typical cumulus as having a roughly pyramidal shape, and if the horizontal movement of the air is small, the loftiest point of the cloud will be situated approximately above the centre of its base. But if the horizontal movement increases in velocity, so that the top is in a more rapidly moving stratum than the base, it will lean forward in the direction of movement. This is a very common phenomenon, being generally shown by cumulus on a windy day.

On much rarer occasions the converse occurs, and the top of the cloud lags behind the base, the explanation being a lessening of the velocity of the wind as the height above ground increases. But such conditions rarely occur, and when they do they are due to local eddies and affect only a limited area. Hence such clouds are isolated, and indicate a disturbed state of the air and uncertainty of weather. The clouds which lean forward are formed under conditions which are spread over wide districts, such as the rear of a large cyclone, and cumulus of that kind may follow one another across the sky for hours or even days as long as the wind persists.

So far we have considered only the round-topped types of cumulus—those which mark the tops of ascending currents whose ascent has been stopped at a comparatively early stage, or those whose ascent is still in that early stage, though the upward movement has not yet come to an end. The full story of the growth of a cumulus is identical with that of the youth of a cumulo-nimbus, the later stages of which we will consider in another chapter.

CHAPTER VII

CUMULO-NIMBUS

GRANDEST of all clouds are the huge mountains of vapour which are the parents of summer thunder-storms. They are at once distinguished from ordinary cumulus by their upper parts, which sometimes reach beyond the region of the alto clouds high into the realm of cirrus, and extend outwards as a broad disc, which is occasionally indistinguishable from the cirro-nebula and cirro-stratus which form the van of a cyclone cloud canopy. Indeed, there seems to be no essential dividing line between a large cumulo-nimbus and the cloud pile of a small cyclone, and no real difference between them except their size.

As a matter of fact, the term cumulo-nimbus would only be given to the cloud when a large fraction of the whole can be seen at once.

In dealing with common cumulus, it has been pointed out that the cessation of the uprising convection currents which determines the maximum height to which the cloud will grow is due to the rate of cooling within the ascending column being greater than the rate of cooling outside it. It follows that when the ascending current has reached a certain height it will, as a whole, be just as heavy as an equal column outside. Ascent must then cease. The equilibrium of the air in such a case is said to be stable, and the condition of such stability is simply that the general rate of fall of temperature per 100 metres of ascent is less than the rate of cooling dynamically produced in an ascending current.

If, however, the general rate of fall of temperature is greater than that produced dynamically, the consequence will be that the upward tendency of the rising air will increase as it moves upward, and the taller the column becomes the greater will be the difference of weight between the inside and outside columns. In such a case the equilibrium is said to be unstable, and the result will be the production of cumulo-nimbus.

Just as cumulus may be divided into heat cumulus and the clouds of the rear of a cyclone, so cumulo-nimbus may be divided into the same two groups. In the case of the heat thunder-clouds the instability of the air is effected by the rapid heating of its lower layers in contact with the ground, those lower layers being so quickly warmed that there is not time for them to become mixed with the overlying air in which the rate of decrease is normal. If there is much wind we rarely get cumulo-nimbus, because the heated air is mixed mechanically with the overlying parts, and the rate of decrease is approximately normal throughout. Calm air and hot sun are then one set of necessary conditions for the production of instability.

But it is well known that thunder-showers and lesser examples of cumulo-nimbus are by no means infrequent in the rear of a cyclone, and such storm clouds are usually attended by considerable wind. They are, as a rule, much smaller than those produced by heat, but they have the same form, and are evidently due to instability in the lower part of the air, and the question is how can that condition be produced. In order to find the answer it is necessary to refer to the temperature phenomena of a cyclonic area. If a cyclone be divided into four quadrants by two lines drawn through its centre, one in the direction in which the system is travelling and the other at right angles to it, then the front right-hand quadrant is the warmest and the rear right-hand quadrant much colder. The cumulo-nimbus clouds of a cyclone are limited to the first part of this cold quadrant, that is to say, to the portion of the storm in which a great volume of cold air is flowing over a district which has just been warmed and wetted by the preceding part. The result is that, the air being warmed by contact with damp ground at a temperature many degrees above that of the air itself, we have produced exactly the same unstable state at a low temperature as we have at a high temperature in the case of heat storms. The lower temperature of the whole is enough to account for the smaller volume of the cloud, and that in turn explains why cyclone thunderstorms are, generally speaking, on a much smaller scale than heat storms.

The life history of a cumulo-nimbus is easily studied on a suitable day. The rapid heating of the lower layers of air causes them to expand bodily, and as they do so they lift the overlying air, frequently in broad domes or waves. The first result is the expansion of these upper zones, which are lightened by the flowing away of still higher layers. Expansion means chilling, and sooner or later its effects become visible in the formation of cirro-stratus, cirro-cumulus, or alto-cumulus. Simultaneously the heated air near the ground begins to rise up in tall columns, while the cooler air from a little higher descends to take its place. Soon patches of lower cloud appear, at first hazy and indistinct, but gradually shaping themselves into cumulus with hazy base and rounded summits. These rapidly assume the typical pyramidal shape, with level base and sharply contoured top, and so far there is little to distinguish them from an ordinary cumulus (see Plate 48). But watch them carefully. Here and there some will be growing taller than their fellows, and as they grow their rate of growth increases until the top begins to show signs of spreading outwards. Rapidly the bulging summit throws out long fingers of cloud, radiating from the central column almost as if propelled by some repulsive force. At first, these fingers are merely projecting lumps of cloud with rounded ends, but in a few minutes they undergo a sudden and striking change. The whole summit becomes frayed out, drawn out into long radiating lines, which thin off against the blue sky exactly like the edges of a sheet of cirro-stratus. False cirrus is the name commonly given to this, but there seems no valid reason why it should be regarded as "false." The top of the cloud rapidly spreads horizontally, forming a disc of cirriform cloud, which sometimes spreads several miles ahead of the rest of the storm. Meanwhile, the original cumulus column loses all its deep folds and convolutions, and other round-topped cumulus arise around it until the completed system consists of a more or less disc-shaped mass of cumulus, with a common base, rising higher and higher towards some central point, where these are connected, by an uprushing column of vapour, to an upper disc with cirriform margins.

Plate 48.

THUNDER-CLOUDS FORMING.

In Plate 49 we have on the left hand a specimen in which the outspreading is just beginning, and the same cloud is shown half an hour later on the left of plate 50. A complete cumulo-nimbus in full work is shown on the right of Plate 49, and the same appears on the right in Plate 50.

Plate 49.
THUNDER-CLOUDS. (CUMULO-NIMBUS.)

Plate 50.
THUNDER-CLOUDS. (CUMULO-NIMBUS.)

These clouds were thunder-clouds, the larger one being a smart thunderstorm with heavy hail. They were photographed in the evening, and in the second picture the sun was just below the horizon.

But, to continue the story of a thunder-cloud, we always find that after a time the cirriform top flattens out and gradually subsides, and this is usually accompanied by a descent of the cloud base to a lower level. Meanwhile, it frequently happens that the whole series of phenomena is repeated in one of the attendant cumulus. Plates 51 and 52 are also two views of the same cloud at different times. In Plate 51 we have the main part of the storm on the right, while on the extreme left a lower part of the cloud is rising rapidly into a tall dome. In Plate 52 the central top has lost its cirriform margin and has distinctly flattened, while the left-hand dome has risen much higher and is beginning to throw out the projecting bits.

Plate 51.

THUNDER-CLOUD. (CUMULO-NIMBUS.)

Plate 52.

THUNDER-CLOUD. (CUMULO-NIMBUS.)

The hard-topped cumulus which fringe the lower disc, and the vast pile of cirriform and hazy cloud which forms the centre of a cumulo-nimbus, are shown in Plate 53, which represents part of the side of a great thunder-cloud. In this case the diameter of the lower disc was about 15 miles, and the upper disc was rather larger. The uprising column in the middle was about 7 miles across, and the height from base to summit about 3 miles. The whole system contained between 100 and 150 cubic miles of cloud. When photographed it was over the northern part of Salisbury Plain. Lightning played repeatedly between the back of the white cumulus and the hazy mass behind it, and the rumble of thunder was all but continuous for nearly half an hour as the great cloud passed by.

Plate 53.

THE FLANK OF A GREAT STORM.

This was an unusually large cloud for this country, but specimens of 10, 20, or 30 cubic miles are quite common.

Now for the explanation of the series of events. To begin with, we have the production of an ordinary cumulus, but the equilibrium is unstable, the growth of the cloud, therefore, becomes more and more rapid, and the rapid condensation adds to the instability until the rising column is so much lighter than an equal column outside that a powerful updraught is created, strong enough for a time to hold up the raindrops or even hailstones. At length the condensation is complete, the upper part of the cloud consisting of snow crystals exactly like those of any other cirrus. In the mean time, the rapid ascending current necessarily involves an indraught from around, and consequent descending currents to supply it. The result is to set up a circulating system, moving inwards along the ground, upwards in the central column, and outwards in the upper disc. The downward currents are sometimes shown by a curling over of the edges of the upper disc, but the phenomenon is not often seen, as the descending movement is generally enough to dry up the cloud particles.

The rapid rising of damp air drawn from the ground brings about rapid condensation and heavy rain. The large size of thunder-drops is almost certainly due to the fact that it is only the larger drops which can fall in the teeth of the strong updraught. But when these drops begin to fall, and still more when cold hailstones begin to fall through this ascending air, it becomes chilled from top to bottom, and the column is broken or even stopped altogether. The frozen particles which make up the top subside gradually, and the chilling of the air immediately below the cloud brings the saturation level nearer the ground, and we say the cloud base descends.

The arrangement of a thunder-cloud into the upper and lower discs with a connecting uprush gives to the typical cloud a shape something like that of an anvil when seen sideways, but in the larger clouds the disc-like form is more obvious. In any ordinary thunderstorm the great majority of the discharges of lightning play between the two discs, and the larger the cloud the more frequent these are. Such discharges as pass between the cloud and the earth come exclusively from the base of the lower disc if the cloud is large, and generally follow immediately after or simultaneously with one between the two discs. The phenomena of lightning are intensely interesting, but the purpose of these pages is confined to the study of clouds and cloud forms, and it would be going beyond our scope to discuss either lightning or hail. Both are, however, so closely related to cumulo-nimbus that they can hardly be passed over in silence. One thing is certain, and that is that neither the electrical developments nor the hail has anything to do with the growth of the cloud. On the contrary, both are consequences of the cloud, the hail being due to the great altitude, and consequent low temperature of the upper part of the cloud, and also to the violent uprising currents within it; while the electrical phenomena are due to either the enormous amount of condensation, or to friction due to the rapid uprush, or more probably to the fact that considerable differences of electrical condition exist in the distant parts of the air connected by the cloud, and between which its circulating currents move. These differences are known to exist at all times, and we cannot here discuss their origin.

The formation of cumulo-nimbus and cumulus is dependent upon the presence of a large amount of water vapour. It is worth while to consider whether the atmospheric movements which bring about the condensation could exist without moisture. Wherever we find differences of temperature between neighbouring places we must get currents of hot air rising from the warmer spots, and compensating descending currents around them. But we have pointed out that if the rate of cooling as we ascend in the still air is less than the rate at which an ascending current will be dynamically cooled, such a rising current will come to rest. If, on the other hand, the rate of cooling in the ascending current be less than in the still air, the equilibrium will be unstable, and a violent uprush will result.

Now, in a climate such as our own, where the lower regions of the air contain large quantities of water vapour, any considerable rise brings about more or less condensation, and that condensation is attended by a liberation of very large quantities of heat, which retard cooling in the ascending current, and so facilitate the production of instability. But if this cause is put aside it is still possible to have a similar circulation. When discussing the causes of instability, it has been pointed out that the prime condition was an unusually rapid rate of fall of temperature in still air, such as may be produced by hot sunshine. Now, these conditions are exactly those which will give rise to the phenomenon of the mirage, and which reach their fullest development in great desert districts when the air is still.

Again, it has been pointed out that the causes which bring showers and thunderstorms to an end include the chilling of the lower parts of the ascending column by the descent of cold rain or hail from above. We may also add the shading of the underlying ground by the cloud itself, and the absorption of heat in the partial evaporation of some of the rain during the lower part of its fall. In a desert district the arising currents are so dry that even a very great ascent does not often result in visible cloud; and when it does, the cloud is produced at so great a height that the air is too rarefied to produce anything much denser than thin alto-stratus, from which no falling droplets could reach the earth. It seems, then, that there will be no such automatic check on the growth of the circulating system, and it will go on growing in volume and intensity indefinitely. As a matter of fact, this is not the case. A different check does come into operation, but not until the indraught and updraught have become so powerful as to draw up the dust and sand and generate a sandstorm, the weight and shade of which, in time, destroys the circulating currents which uplifted them.

Since, however, condensation is a considerable factor in producing instability, we should expect that such sandstorms would be rarer than thunderstorms are in an equally hot but well-watered district, which is the fact. Again, since rain and cloud are checks upon such systems, we should expect the sandstorm systems to be larger and far loftier than thunderstorms, and to consist of far more violent atmospheric movements. This also is the case, and when we know that some of these disturbances have the dimensions of a cyclonic storm, it is easy to understand how the finest dust may be raised to vast altitudes, into the great upper currents of the air, by which it may be borne hundreds of miles before returning to the ground. It is thus that the dust of the African deserts is carried across the Mediterranean to Europe, and the yellow loess from Mongolia even to the eastward of Japan.

CHAPTER VIII

WAVE CLOUDS

REFERENCE has already been made on more than one occasion to the remarkable rippled or wavy structure sometimes assumed by clouds. The waves may be of almost any dimensions, from the broad bands into which a sheet of cirro-stratus or of alto-stratus is sometimes divided, down to the most minute ripples. Sometimes they are ranged in long straight lines, sometimes they are bent into sharp angles, and sometimes curved in very elaborate patterns; but whether they be large or small, straight or curved, no one can see them and fail to conclude that they must be due to an action more or less analogous to the causes which produce waves on the sea or ripple marks upon the sand.

Wave clouds occur at all heights where clouds are formed. The break-up of a lifting fog into roller clouds is probably the lowest example, but it may more frequently be seen in higher clouds of the alto or cirrus kinds.

A low example is given in Plate 40, which represents stratus maculosus, and which has already been described. A higher type is shown in Plate 54, which is a wave-like arrangement of alto-cumulus. Rather higher come the long zig-zag bands of Plate 55, in which the stratiform arrangement is more obvious, and which would be best described as a wave-form of alto-stratus. These two plates form striking contrasts. The clouds shown in the first are distinctly of the cumulus order, and a prominent feature is the way in which the right-hand side of each wave has a clear-cut rounded contour like that of the upper edge of a small cumulus, while the left-hand edge of each band is frayed out into a ragged fringe. The whole cloud was moving slowly in a direction nearly, but not quite, at right angles to the waves, and the fringed edge formed the rear. It is evident that this peculiar structure must be due to a series of narrow waves intersecting a plane in which the air is just on the point of producing alto-cumulus. If there were no such waves, the little uprising currents, with their intervening down currents, would be irregularly distributed, and all the wave disturbances have had to do is to arrange them. The consequence is that as the waves pass along the stratum the air is alternately raised and lowered. Where it is rising condensation takes place, where it is falling evaporation results.

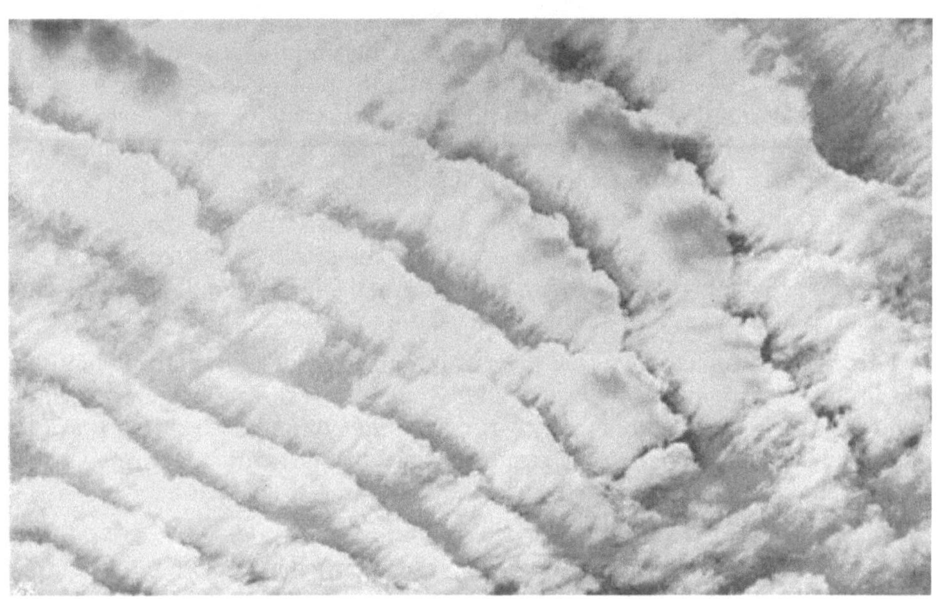

Plate 54.

CRESTED ALTO WAVES. (ALTO-CUMULUS UNDATUS.)

Plate 55.

ALTO WAVES. (ALTO-STRATUS UNDATUS.)

The cloud, like most other wave clouds, did not retain its features for any length of time, but the gaps closed slowly in as the cloud-bands increased in size, until a sheet of alto-stratus was produced. Since the time of day was the morning, it is almost certain that the plane of saturation was rising in accordance with the general law, which is that the planes of condensation rise steadily, until about two or three o'clock in the afternoon, and then slowly descend.

In Plate 55 each band is much flatter and less dense. They are just as evidently formed by wave movements intersecting the plane of condensation; but this was formed in the evening when the sun was nearing the horizon, and at a time when the cloud planes are as a rule rapidly descending.

Among the alto clouds wave-forms sometimes persist for a fairly long time, and in this case the bands moved steadily onward in a direction equally inclined to their length and breadth, that is to say, from the bottom left-hand corner of the photograph to the top right-hand corner. As they passed across the sky new bands kept on making their appearance at about the same spot, each band persisting with little change until it had passed out of sight.

Going much higher up into the region of cirrus, we meet with the most minute and delicate ripple clouds. Some of these have already been referred to. They are connected with either cirro-macula, cirro-cumulus, or cirro-stratus, just as the coarser textured waves we have been considering are connected with alto-cumulus or alto-stratus. In Plate 56 we have an example in which we can see the stages in the process. Nearest to the zenith we have cirro-cumulus, which is here and there irregularly distributed, but is generally arranged in delicate ripples, which are variously curved. Nearer the horizon the troughs of the waves are filled in, and sheets of cirro-stratus are the result. Here, again, the wave-form is evidently not typical. It is an arrangement of either cirro-cumulus or cirro-stratus, produced by the intersection of the plane of condensation by a series of wave movements.

Plate 56.

CIRRO RIPPLES. (CIRRO-CUMULUS UNDATUS.)

The arrangement is, however, so striking a feature when it is well shown that any description of the cloud which contains no reference to the waves is manifestly incomplete, and this would be best effected by adding the word undatus or waved to the name of the cloud. Plate 54 will then be alto-cumulus undatus, Plate 55 alto-stratus undatus, and Plate 56 would be described as cirro-cumulus undatus, passing into cirro-stratus undatus and cirro-stratus. In popular language Plate 55 might be called alto waves, Plate 54 crested alto waves, and Plate 56 cirro ripples.

If we are satisfied that the wave clouds are due to a wave movement intersecting a plane of incipient cloud formation, the whole question of their mode of production resolves itself into two parts—how is that plane of incipient condensation produced? and how can we account for the intersecting waves?

The first question has by far the greater importance, since it amounts to asking for a general explanation of the production of high clouds, especially the forms of cirro-cumulus, cirro-stratus, cirro-macula, and the corresponding alto varieties. There are, again, two divisions also to this question. How does the water vapour reach the stratum in sufficient quantity to saturate it? and when condensation takes place, why does it so frequently assume the characteristic mottled and granular forms like crowds of little cumulus clouds arranged in one level? This last sentence gives the clue. They are, in truth, little cumulus clouds, and must be formed in exactly the same way as their vastly larger prototypes of lower regions. It has been explained that low cumulus is the result of large upward moving air columns or convection currents, each one being initially caused by the heating of the vapour-laden air near the ground, and each uprising column being supplied by cooler descending air which flows down in the intervening spaces. It has also been explained that these movements result in changes of temperature, which tend to check those movements and restore the original equilibrium. Suppose this to occur, as it constantly does, without any column reaching sufficiently high to produce a cloud. There will be no visible effect, but, nevertheless, an important change has taken place. Every ascending current has lifted some water vapour with it to a higher level, and the descending drier air has come down in contact with the ground or damper air to become equally charged with moisture in its turn. The process will be repeated again and again, and at one level after another, so that the water vapour travels ever higher and higher.

This process of interchange between ascending and descending air has been called by Mr. Ley inversion, but the term does not seem very suitable, and inter-convection would be better. The two opposite currents pass through each other, as if the ascending air gathered itself into definite channels, and passed through holes in the descending mass like the passage of water upwards through a descending plate of perforated metal. Moreover, just as the holes in such a descending plate might have any size, so that the ascending streams might vary in breadth from the finest hair to a column of huge diameter, in exactly the same way the ascending columns of air may vary from the smallest imaginable size to the great cumulo-nimbus currents. It is the little currents which account for the constant quiver of the margins of any object which is viewed through a large telescope by day, and for the haze, so characteristic of a hot day, which makes distant objects seem ill-defined and in a state of continual tremble. The rays of light in passing through the intersecting streams are bent a little, now this way, now that, as the air currents sway to and fro.

The near neighbourhood of the ground is not essential. As long as the temperature of the air at any level is rising, so long interconvection must occur. The process will be independent of the presence or absence of wind. All that wind can do is to mix up the air at different levels, breaking the system of currents and reducing it to, so to say, a finer texture, or producing eddies, if strong enough, which direct the currents and gather them into definite channels. The final result in any case is that, with rising temperature, water vapour is steadily borne upwards from the ground.

As it ascends the air becomes cooler, and yet retains its water vapour. When the rising currents are large they mix little with the descending dry air, and on reaching a certain level condensation takes place, and we have the beginning of a cumulus. If they are of a more moderate size they will ascend less rapidly, the admixture with descending air will bear a larger proportion to the whole, and the plane at which condensation will begin will be higher, and then each small column will be tipped with a ball of alto-cumulus. Make the interconvection currents smaller still, and the cloud plane will be lifted yet higher, and we shall have cirro-cumulus or cirro-macula.

Now, the more even the distribution of temperature on the ground the less the probability of coarse interconvection, and the same is true of any higher stratum of air, provided it is free from disturbing influences from outside. If, therefore, we have large currents near the ground, ending, as they must, in cumulus, it has already been explained that these clouds stop the action, and the general system of large currents will be restricted to the region in which they occur. At some distance above the lower clouds the only difference will be that water vapour has been brought up to their level in great abundance. Smaller systems of interconvection can then exist, and so we may have the spectacle of several layers of cloud—cumulus capping the great currents of lower regions, alto-cumulus forming the summits of the smaller currents of intermediate regions, and cirro-cumulus floating far above both.

Frequently it happens that before the ascent of vapour has gone quite far enough to produce a cloud, other causes co-operate, and the cloud makes its appearance suddenly over considerable patches of sky. The most potent of these is a fall of the barometric pressure, which is brought about by some of the air far above the region of even the highest clouds flowing away to some other district. The air at all lower levels being thus relieved of the superincumbent pressure, immediately expands, and is thereby cooled throughout. Consequently, if at any level it was near its point of saturation, it will be carried beyond that point, and cloud will rapidly make its appearance over a large part of the sky, possibly at more than one level. Stratiform arrangements will be the rule; but if interconvection is going on at the time, its presence will be betrayed by a granular or cumuloid structure. Interconvection clouds should then be most frequent, and best formed when the air as a whole is still or moving slowly (so as not to create great eddies), when the temperature is rising rapidly, and when the barometer is making a sudden fall. All these conditions are met in thunder weather, and at the time when a summer anticyclone is giving way. It will be remembered that many of the most beautiful forms have been described as forming under one or the other of these very conditions.

A second contributing cause, and one which tends to make the condensation in patches or long broad bands ranged roughly at right angles to the direction in which the air is moving, has been referred to earlier. It is the passage of the air over an undulating country; the up-and-down movements of the lower air being transmitted upwards to great altitudes, as ever broadening and flattening waves. If the upper air is flowing more rapidly than the lower, these broad waves may be far ahead of their real cause, which will, therefore, quite escape recognition, but the phenomenon is constantly to be detected in the arrangement of the lower clouds. Two instances in the writer's experience will suffice. It was desired one morning to measure the altitude of some small clouds which were passing from the north-west at a height of probably between 2000 and 4000 metres, over a hill only about 150 metres higher than the valley in which the apparatus was fixed. In order to make the measurement, it was necessary for the cloud to cross the valley and appear in the same field of view as the sun, according to the method that will be described further on. But in order to cross the valley the air had to descend, and so, of course, had the cloud stratum, though to a less extent. But small as the descent was, it was enough to dry up the clouds entirely, and for more than a couple of hours the clouds came sailing over the hill, disappearing entirely, and then reforming so far beyond that no measurement was possible, since not one single fragment came near enough to the position of the sun, which remained shining brightly through a broad clear gap between two patches of cloud-strewn sky.

On another occasion considerable preparations had been made for some photographic observations during an eclipse of the sun. The observatory stands on the eastern side of the valley of the Exe, which is flanked on its western side by a long ridge of hills going up to 800 feet above the sea. Beyond these hills lies the deep, narrow valley of the Teign, and beyond that the granite ramparts of Dartmoor, 1000 feet above the sea. The wind was blowing gently across the two valleys, and shortly before the eclipse began a broad strip of thin cloud formed above and rather towards the eastern side of the Exe valley, just where the sun was, while at the same time the sky was practically clear half a mile further east, and bright sunlight was streaming down on the ridge between the two rivers a few miles towards the west. The cloud was never thick enough to quite hide the sun, so that the eclipse was easy to watch with the naked eye; but in spite of fairly rapid movement of the cloud masses as they drifted before the sun, they kept on forming in just the same place, and completely prevented the carrying out of the programme planned. It is almost certain that the phenomenon was brought about by an upward moving wave marking the place where the level of approaching saturation was upheaved by the disturbance caused by crossing the two valleys and intervening ridge.

These two instances are not quoted as examples of a rare occurrence, but as definite simple instances of a phenomenon which may be constantly observed, and as proof that the conformation of the ground does exercise an influence upon the distribution of cloud.

But no irregularities of the ground will suffice to explain the minute waves and ripples which have been described at the beginning of this chapter. These must be due to wave disturbances in the air itself. They have been explained as due to two different currents of air, either a warm damp current flowing over a cold one, or *vice versâ*. Now, such an occurrence as a warm damp current flowing over a cold one must be very rare, though it is impossible to deny that it might occur. The immediate contact of a cold current above a warm damp one is equally unlikely, unless the general atmospheric condition were greatly disturbed, which is the same thing as saying that wave clouds would not occur. They are most frequent at just those times when interconvection has freest play, and this is amply sufficient to account for a plane of saturation without any necessity for a hypothesis of two layers of air at different temperatures all but producing cloud at their junction. No convincing evidence of cloud production by such means has yet been adduced, and it is better to rely upon causes which we know do operate than to call in theories as to what might possibly happen. This is one of those points in the study of clouds which need investigation, and until proof is forthcoming it is better to say that the admixture of two strata of air might conceivably produce cloud, but most forms can be accounted for by other causes of which we have more positive evidence.

Still, the wave clouds are due to waves, and there seems no other way of accounting for them than the supposition of gentle differential currents. But if such currents occur the ripples and waves will not be limited to a definite surface, so to say, of contact, but will be propagated upwards and downwards for considerable distances from the level of greatest disturbance. Whether, therefore, the level at which the natural operation of interconvection has produced saturation is high or low in this region, the result will be the marshalling of the ascending and descending elements of the convection system in the characteristic waves.

The differential currents, then, which cause the waves must not be conceived as producing those waves at a surface of contact, nor must the currents be thought of as separated by any definite surface, but rather by a region of variable but usually considerable depth, in which the air is disturbed by a series of small slow eddies and oscillatory movements. When the waves are parallel straight lines the air currents may be really portions of a whole, having the upper part more rapid than the lower. In such a case the direction of movement should be at right angles to the cloud lines. If the upper current differs in direction as well as velocity, the direction of movement of the clouds will be intermediate, and will resemble that of the upper or lower current, according to their relative distances from the plane at which the clouds are formed.

The behaviour of the clouds will depend upon the relative shares in their production borne by interconvection pure and simple and by the wave oscillations. If the stratum is one in which cloud would actually be formed independently of the up-and-down movements, all this will be able to do will be to arrange the cloudlets at their birth, and these will then continue to exist, drifting with the general horizontal movement of the air like any other cloud of the same order.

On the other hand, if the production of cloud is dependent upon the vertical oscillations, the cloudlets or lines of cloud will move with the air waves, and their rate of motion and direction of motion will be determined by the rate and direction of the waves, which may be quite different from that of the air at that stratum as a whole. The ascending waves will be marked by lines of cloud generally rounder and better defined on their advancing sides, while the descending troughs will be marked by clear intervals.

Wave movements of the necessary kind are frequently very complicated, and it is not by any means a rare occurrence to see the wave lines in one part of the sky at all sorts of angles with similar lines in other parts, or even to see two or more sets of waves at different altitudes crossing one another. Either phenomenon is always accompanied by rapid changes in the cloud, and the rippled structure is short-lived. This was the case with the clouds shown in Plate 54. Plate 53, on the contrary, shows great uniformity in the wave lines, and although the vertical oscillation is probably the main cause of condensation, the form was unusually persistent.

Irregular patches of wave disturbance, affecting a plane occupied by cirro-stratus vittatus, are shown in Plate 57. In this case the wave systems only touch the cloud plane here and there, and the places of contact varied rapidly. It is pretty clear from this photograph that the idea of the waves being formed at a surface of contact between two diverse currents will not suffice. The bands of the cirro-stratus are for the most part unbroken and unaffected; it is only here and there that the wave region touches them.

Plate 57.

WAVED CIRRO-STRATUS. (CIRRO-STRATUS UNDATUS.)

The conclusions at which we have arrived are simple, and there is little room for doubt as to their main correctness, but there are numerous minute features presented by these beautiful cloud patterns which await interpretation, and they reveal complicated oscillatory movements in the air which are difficult to account for, whether we seek their originating causes or the mechanics of their motions.

CHAPTER IX
CLOUD ALTITUDES

DURING an extended experience of cloud photography, it was found that it was quite possible to get pictures which showed the cloud detail even when the sun was in the field of view. Sometimes the solar image was reversed, but if the exposure was very short this was not the case. In such photographs the structure of the cloud was exceedingly clear and sharply defined quite close to the sun. Indeed, the intense illumination seemed to reveal minute details of internal arrangement which could not be detected in similar clouds some distance away.

The methods which had been employed for the measurement of cloud altitudes elsewhere have already been briefly referred to. Some of them required two observers, who were equally responsible, each of them having to direct his apparatus or camera to the same point of the cloud, and to record the exact direction in which the instrument was pointed. The instruments, if accurate, were costly, and there were many opportunities for error in reading the graduated circles which gave the directions. Moreover, in most of these methods the two observers were connected by telephone, and had to agree on the exact point towards which their instruments should be directed; either the exact point of the cloud, or the precise direction as shown by the mounting of the camera or other instrument. At Kew some of these sources of error were avoided by fixing the two cameras with the axes of the lenses and centres of the plates in a vertical position and exposing the two plates simultaneously. The Kew observations were not long continued, and for some years the only measurements in progress were those carried out abroad, particularly at the Blue Hill Observatory and at Upsala.

The experience gained in photographing clouds in order to record their forms suggested a way in which many of the sources of error in previous measurements of altitude could be avoided, especially by simplifying and reducing the operations at the moment of making the observation.

If two cameras are placed at the opposite ends of a measured base line, whose direction is known, and if they are both pointed towards the sun, on making the exposures by electrical means at the same moment, the position of the image of the sun upon the plate gives the direction in which the cameras are pointed. It will be in the same direction as seen from both ends of the line.

Now, if we note the time at which the exposure is made, this with the date gives all that is required for ascertaining the sun's position in the sky, and is, therefore, the only exact observation which need be made at the time of taking the photographs. Mistakes are almost impossible, as each plate contains its own record of the sun's position, and even if some of the plates should get mixed the images of the clouds will generally suffice to pair them properly. For general measurements there is one grave defect in the method, and that is that it can only be used when the sun and cloud can be got into the same field of view. But with the higher varieties of cloud this is generally possible, and it was just these higher sorts about which knowledge was least certain, and which it was proposed to study.

An initial difficulty was the finding of a level site, flat land being very uncommon in Devonshire, but fortunately a suitable place was found in some artificially levelled ground close to Exeter, belonging to the London and South Western Railway Company. It was a stretch of ground intended to be covered with sidings, but had not been finished, and had become overgrown with grass, stunted sallows, and other wild plants. Being railway ground, it was, comparatively, though by no means entirely, free from mischievous and inquisitive people. The next point was a suitable camera. It must have fairly long focus in order to give a large image, and therefore large displacement; it must be capable of being pointed in any direction and clamped there; and it must be capable of standing considerable extremes of temperature and variations of dampness, as it was intended that they should be kept on the spot in wooden structures, which served for stands as well as to contain the apparatus.

The pattern finally decided upon is represented in Plate 58, which shows one of the cameras pointed up to the sky and standing on one of the stands. These cameras were to take plates of whole plate size, two double dark slides of the ordinary pattern being attached to each.

Plate 58.

CAMERA FOR MEASURING ALTITUDES.

The camera looks rather complicated, but it is really simple. Its body consists of front and back, each attached to a central part by a short bellows and sliding on a base board, to which it can be clamped by screws of the usual pattern. The central part carries trunnions, such as are used for looking-glasses, which swing in sockets carried by two upright supports, so as to give the whole free motion in a vertical plane. In order to be able to fix it firmly at any angle, the base board of the camera body carries on its underside a thin board projecting beneath it and forming a segment of a circle whose centre would be the horizontal axis through the trunnions. The board passes between the jaws of a small wooden clamping vice in front, which is carried by the square base to which the uprights are fixed. The whole is firmly made of well-seasoned pine, and has stood well the hard usage of half a dozen years.

There is no focusing screen. Focusing was done with great care once for all, and then a coat of hard varnish was put over all the adjusting screws. A small view-finder is attached to one side, and it was by this that the camera was pointed in the desired direction.

In order to lessen risk of mistake, it was so arranged that the two slides belonging to one camera would not fit the other. The lenses, of 18 inches focus, and giving sharp detail all over the plate, were carefully matched, and the focus adjusted until the images given by them when placed side by side appeared to coincide exactly. They were provided with iris diaphragms, which were shut down to an aperture of a quarter of an inch, and with shutters which could be released at the same moment by an electric current, acting through the electro-magnet shown under the lens on the front of the camera.

The shutters were of the kind known as the "Chronolux," which will give any exposure from the sixty-fourth of a second up to three seconds. But it was found in practice that the highest speed was sufficient and gave satisfactory results. Of course, there was no idea of adjusting matters on each occasion so as to get the best possible negatives capable of yielding good prints. Measurement was the object, and if the negative showed the sun and sufficient cloud detail for the identification of cloud points, that was all that was wanted. The shutters gave a good deal of trouble at first. Their sliding parts were made of ebonite, and when the cameras were left in their stands with an August sun shining down upon them, everything inside got very hot and the ebonite warped; but the difficulty was got over by substituting aluminium.

The two camera stands were placed 200 yards apart, and were connected by a line of telegraph wire carried on short poles. At each end of the wire an insulated connecting piece was brought down to the camera stand, and to the batteries and other apparatus. The current which was sent through this wire by pressing a contact at one end of the line did not directly make the exposures; but two similar relays were brought into action, and each of these sent the current from a local battery of Leclanché cells through the electro-magnet on the camera and made the exposure.

After development the two negatives showed the image of the sun, not far from the centre of the field of view, and the cloud whose altitude was required. Since this was taken from two different points of view, the negatives were not alike, but the distances between the centre of the sun's disc and any special point of the cloud were different. For instance, if the cloud were east of the sun, with its edge just apparently touching the solar image as photographed from the eastern station, then the negative taken from the western end of the base would show an interval of clear sky between the two, which would be greater as the cloud was lower.

It often happened that after developing the plates the image of the sun was lost in a black blur, but it was easy to reduce this part of the image by local application of a reducing agent[3] by means of a paint-brush, until the disc became clear enough. Two lines were then drawn on the negative, one vertical and the other horizontal, intersecting each other at the centre of the sun's image. These lines served as the starting-points for exactly measuring the distance from their point of intersection to any selected point of the cloud.

The distances could generally be determined to a fiftieth or a hundredth part of an inch, and their difference was, of course, dependent upon the direction of the sun relative to the base line and the altitude of the cloud, but for low level clouds the difference was sometimes so great that no pair of corresponding points could be detected, while it was often as much as an inch. With higher clouds the differences were smaller, but unless the sun was very low in the sky, either east or west, the displacements of the cloud image were great enough to give reliable measures. Specimen prints from pairs of negatives are shown in Plates 59 and 60.

---

[3] Ferricyanide of potassium and hyposulphite of soda.

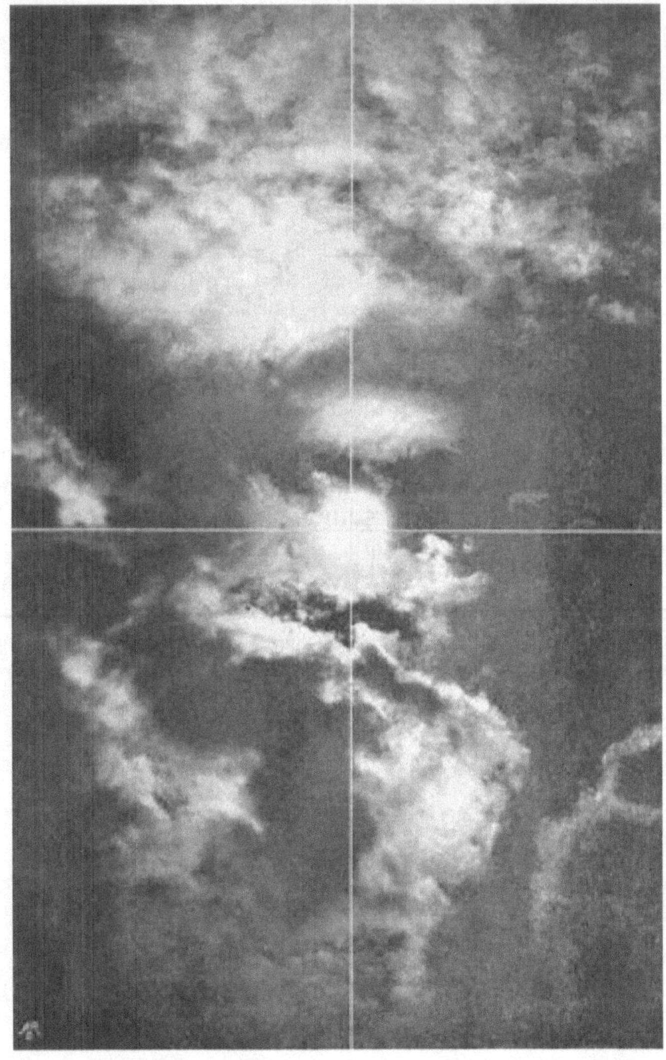

Plate 59.

PRINT FROM A NEGATIVE USED FOR MEASURING ALTITUDE.

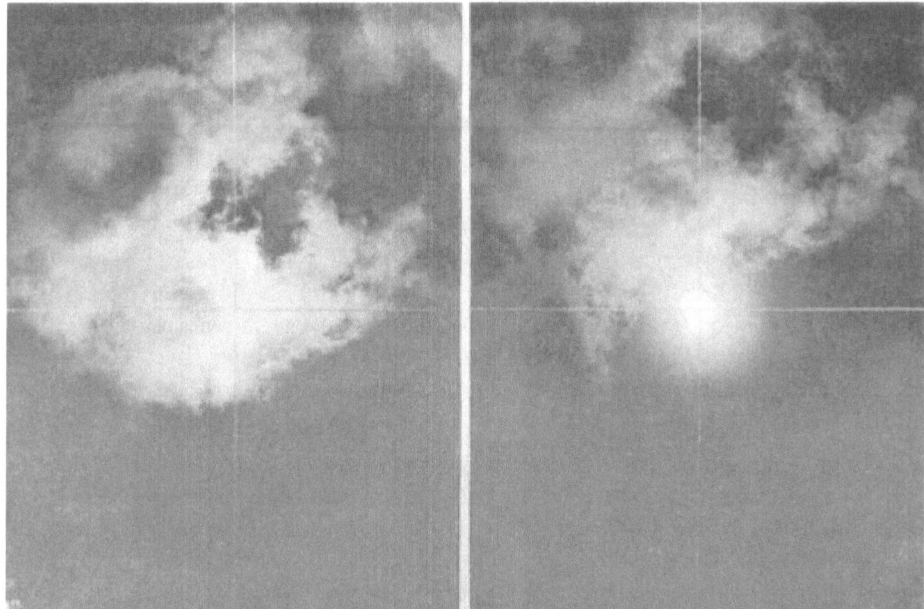

Plate 60.

PAIR OF PRINTS SHOWING THE DISPLACEMENT OF THE CLOUD.

The processes by which the measurements are worked out are laborious,[4]

[4] From the declination of the sun corrected for variation and from the known latitude, the meridian zenith distance is calculated.

From the Greenwich time, the longitude, and the equation of time, the hour angle is obtained.

Now, if H be the hour angle, D the reduced declination, and M the meridian zenith distance, the sun's altitude may be calculated by the formula—

$$\log \text{versin } H + L \cos \text{lat.} + L \cos D\text{-}20 = \log n,$$

where n is a natural number, and

$$n + \text{vers } M = \text{covers alt.}$$

Again, to find the azimuth—

$$\text{vers sup. (lat.} + \text{alt.)-vers polar dist.} = m,$$

where m is another natural number, and

$$\log m + L \text{ sec. lat.} + L \text{ sec. alt.-20} = \log \text{vers azim.,}$$

reckoned from the south.

Hence the position of the sun is ascertained for both negatives.

By actual measurements on the plates and reference to a previously constructed scale the position of the cloud as seen from each camera is next determined, and the angle subtended by the base line at a point X vertically beneath the cloud is calculated. If A and B are and consist of two parts, the first being the determination of the exact position of the sun from the date, hour, and latitude and longitude of the place, and the second, the determination of the position of the cloud. Two points which represent the same part of the cloud are selected, and their respective distances from the two lines drawn through the sun are measured as accurately as possible. Now, a certain distance on the negative corresponds with a definite angular displacement, and a scale can be constructed showing how much should be added to or subtracted from the sun's position to get the exact position of the cloud. This being done, it is then a simple piece of trigonometry to deduce the actual height of the cloud above the place of observation. The work of computation, however, was greatly lightened by the fact that many of the pairs of negatives showed more than one layer of cloud; thus Plate 59, which is a fair specimen, shows three layers, and, consequently, one determination of the sun's position sufficed for three distinct results.

For the highest clouds the displacements were, of course, small, and could only be made with certainty of a correct result within about three hours of noon. Earlier than 9 a.m., or later than 3 p.m., the sun was too nearly in a line with the two stations, or too low in the sky, to give a sufficient displacement of image. A base line of 400 yards instead of 200 would have been better for the high clouds. But, on the other hand, when low level clouds are viewed from two different spots their outlines may seem so changed that it may be impossible to identify a pair of corresponding points, and the same difficulty may also arise when high clouds are seen through a gap in a lower stratum. The longer the base line the more frequent and more obtrusive would this perspective difficulty become, so the distance of 200 yards between the stations was adopted as a convenient mean.

The method of making the observations was simple. Each observer was provided with some signal flags, by which the necessary communications were made in accordance with a simple code. Call the two observers A and B, and suppose A directed the operations. He watched the sky until a favourable opportunity seemed to be approaching. He then signalled to B, and both cameras were turned to the sun, the dark slides were inserted, the shutters set, and everything made ready. Signals were then interchanged, to signify that preparations were complete, and when A saw that the edge of the cloud had reached a suitable position to be in the same field of view with the sun, the contact key was pressed and the plates simultaneously exposed. At the moment when this was done the time was noted. Several observations were thus made in a short time.

Measurements were carried out as opportunity allowed over four consecutive seasons, from the beginning of April until the end of October. During the last of the four years, the site had become less convenient owing to an extension of the railway work, and early in November the series was brought to an abrupt conclu-

---

the stations, and a and b the angles from them respectively, the distance AX is given thus—

$$\log AX = L \sin b\text{-}L \sin AXB + \log AB,$$

and the height h of the cloud above X is given by—

$$\log h = \log AX + L \tan \text{alt.-}10.$$

sion by a heavy gale, which snapped off all the poles carrying the connecting wire. But by that time 423 measurements had been obtained, the great majority of which referred to clouds of the cirrus and alto groups.

The general results may be tabulated thus, giving heights in metres:—

	Number of observations.	Maximum altitude.	Minimum altitude.	Mean altitude.
Cirrus	58	27,413	4,114	10,230
Cirro-stratus	64	15,503	3,840	9,540
" cumulus	63	11,679	3,657	8,624
Alto-cumulus	83	9,390	1,828	5,348
Cumulus top	42	4,582	—	3,006
" base	48	1,959	584	1,290
Strato-cumulus	27	6,926	823	2,248
Cumulo-nimbus top	15	6,409	2,004	8,002
" " base	15	2,286	766	1,045

These values are not very different, on the whole, from those which have been arrived at elsewhere, and in making a comparison it must be borne in mind that there is always a little want of precision in cloud nomenclature. As a whole, the Exeter maxima are greater than the foreign ones, and this is very markedly so in the case of cirrus, for which the American highest record is 14,930 metres, the Swedish record is 13,376, while the Exeter value is 27,413 metres, or about 17 miles. But this extreme measurement, and several others unusually large, were made in one morning, a day of very hot damp weather, when cloud formed at seven different levels: cumulus at a height of 1·9 miles, alto-cumulus at 3·9 miles, cirro-cumulus at 4·7 miles, cirro-stratus (No. 1) at 8 miles, cirro-stratus (No. 2) at 9·6 miles, cirrus at 11·5 miles, and cirrus excelsus at 17 miles. By about half-past one in the afternoon the sky was completely overcast with dull grey clouds, which cleared off at half-past four, and at half-past five in the evening the cirrus had fallen to 7·9 miles, and the cirro-cumulus to 4·3 miles. If this one day's observations had been omitted, the Exeter maximum would only have been little more than 1000 metres above the record from across the Atlantic, but 1000 metres is a height worth noting.

While the Exeter maxima are all rather greater, we find the minima for cirrus, cirro-stratus, and cirro-cumulus are rather less than at the foreign stations; that is to say, that clouds are formed over Devonshire both at lower and at higher levels than seems to be the case in Massachusetts or Sweden. It seems probable that this is due to a greater humidity on our western coasts, such as we should suppose would be the case from their position and the prevailing winds and ocean currents. If so, we should expect the great convection clouds to be larger. Thus, at Exeter, out of only fifteen examples of cumulo-nimbus, the top varied from 2004 metres to 6409, with an average base level of 1045. At Upsala the maximum was 5970 and the minimum 1400, with an average base level of 1400. The mean thickness of the Swedish clouds was only 1400 metres, while that of the Devonshire specimens was more than 2000 metres.

Again and again, during the progress of these measurements, it was found that the greatest altitudes and the richest development of the higher varieties occurred towards the end of a spell of fine calm weather, when convection had had free play day after day. A slight fall of the barometer, only the hundredth part of an inch, would usually, under those circumstances, bring about abundant formation of high clouds, frequently of the undatus kind. All the cumulus clouds, by which we mean to include alto-cumulus and cirro-cumulus, are most frequent when the levels of condensation are rising, while the stratiform clouds are an indication of no vertical movement or of active descent. Pure cirrus is indicative rather of movement in a horizontal direction, and may occur when the condensation levels are stationary, or when they are rapidly changing either way.

In broken weather the natural movements of the atmosphere and of its vapour are masked and disturbed by the strong eddies brought by the cyclonic systems. It not unfrequently happens that the region of disturbance does not reach up to the level of the highest cirrus, or, what is more probable, the cyclonic system leans so far forward that we may have in its rear the upper clouds floating quietly far above the comparatively shallow region of disturbance, while in front the upper part of the storm system projects above undisturbed air.

The frequent appearance of cloud almost at the same time at more than one level is at first rather difficult to understand, but it will be noticed that when this occurs the barometer almost invariably falls. Now, if we suppose that the air is nearly saturated at more than one level, and that the whole is then bodily relieved of some of the superincumbent mass, so that the barometer falls, the mass of air will at once swell up, being cooled from top to bottom simultaneously, and wherever it is damp enough cloud will be formed.

The converse is equally true. If we have cloud at several levels, and the whole is compressed by the addition of more air above, which is the case when the barometer rises, that compression will be accompanied by the generation of heat and the consequent disintegration and disappearance of the clouds.

CHAPTER X
CLOUD NOMENCLATURE

SINCE a considerable number of new terms have been suggested in the foregoing pages, it may be convenient to collect them and tabulate them, so as to show their relation to those already recognized by the International system.

In the atlas put forward by the committee, sixteen varieties are recognized by distinct names, and these are drawn up in tabular form with appropriate abbreviations for use in making records.

The names are—

Cirrus. Ci.	Cumulo-nimbus. Cu. N.
Cirro-stratus. Ci. S.	Stratus. S.
Cirro-cumulus. Ci. Cu.	Fracto-cumulus. Fr. Cu.
Alto-cumulus. A. Cu.	Fracto-nimbus. Fr. N.
Alto-stratus. A. S.	Fracto-stratus. Fr. S.
Strato-cumulus. S. Cu.	Stratus-cumuliformis. S. Cf.
Nimbus. N.	Nimbus-cumuliformis. N. Cf.
Cumulus. Cu.	Mammato-cumulus. M. Cu.

During our survey of these groups we have found that some of them include clouds of many shapes, which must be due to very diverse conditions. It follows that if observations are to be made on the occurrence of these special kinds, with a view to arriving at a thorough understanding of the circumstances to which they owe their forms, it becomes necessary to devise a code of names and symbols whereby an interchange of ideas and records may be rendered possible. Specific names have been proposed as each form was considered, and it only remains to sum them up concisely. Subsequent observation, particularly in other climates, may show that further additions should be made; but if the principle of specific names be once admitted, it will be easy to fill any omission.

GROUP CIRRUS.

Under the general head of cirrus we have found nine distinct forms—

1. *Cirro-nebula* (Ley) (Plates 2 and 3). Cirrus veil.

Characterized by comparative absence of structure and by the formation of halo. Ci. Na.

2. *Cirro-filum* (Ley) (Plate 7). Thread cirrus.

Built up of fine long threads, straight, curved, or crossing, but free from hazy curling or flocculent structures. Ci. F.

3. *Cirrus excelsus* (Plate 5). High cirrus.

Characterized by great altitude, thinness, irregular branching structure. Ci. Ex.

4. *Cirrus ventosus* (Plate 6). Windy cirrus.

Characterized by curving branches leaning forward in the direction of movement, and other long curving streamers lagging behind and below. Fluffy parts are usually present, and mark the origins of the long curling fibres. Ci. V.

5. *Cirrus nebulosus* (Plate 9). Hazy cirrus.

Characterized by the absence of sharply defined lines, fibres, or streamers; all parts of the cloud being hazy, and suggestive of other varieties of cirrus out of focus. Ci. Neb.

6. *Cirrus caudatus* (Plate 8). Tailed cirrus.

Characterized by small hazy or fluffy heads behind or below which hang long streamers, which taper away more or less to a point. The tails are sharply defined, and so are the edges of the heads. Ci. Ca.

7. *Cirrus vittatus* (Plates 12 and 13). Ribbon cirrus.

Characterized by formation in long bands of cloud, sometimes made of parallel long fibres with cirrus haze linking them together, sometimes consisting of a long bundle of fibres, from which others diverge at an angle as shown in the plate. Ci. Vt.

8. *Cirrus inconstans* (Plate 10). Change cirrus.

Characterized by a peculiar ragged, wavy appearance. It is generally only the beginning or the end of a mass of cirro-stratus or cirro-cumulus, but occasionally it vanishes shortly after its appearance, without reaching the further stage. Ci. In.

9. *Cirrus communis* (Plate 11). Type cirrus or common cirrus.

Characterized by short irregularly curling fibres collected together in considerable patches. No definite arrangement into any of the forms already described. Ci. Com.

GROUP CIRRO-STRATUS.

Under this group the cloud usually shows some structure, being apparently built up from a massing together of detached forms at a common level. When this is so it should be described by adding the specific name of the detached form most nearly related.

1. *Cirro-stratus nebulosus* (Plates 3, 4, and 14). Hazy cirro-stratus.

Characterized by absence of visible structure. Ci. S. Neb.

2. *Cirro-stratus communis* (Plate 16). Common cirro-stratus.

Characterized by the presence of short curling fibres matted together. Ci. S. Com.

3. *Cirro-stratus vittatus* (Plate 57). Ribboned cirro-stratus.

Characterized by being made up of long stripes or bands of cloud. Ci. S. Vt.

4. *Cirro-stratus cumulosus* (Plate 17). Flocculent cirro-stratus.

Characterized by an obscurely granular structure. Ci. S. Cu.

Many forms of cirro-stratus are arranged in waves or ripples. This is indicated by attaching the word undatus, or waved, after the ordinary specific name, or the letter U after the abbreviation.

GROUP CIRRO-CUMULUS. Divisible into three species.

1. *Cirro-macula* (Ley) (Plate 23). Speckle cloud.

Characterized by semi-transparency, by the fact that the particles are frequently whiter and more opaque on their edges. A patch of cirro-macula always looks like a thin sheet which has curdled. Ci. Ma.

2. *Cirro-cumulus nebulosus* (Plates 20 and 21). Hazy cirro-cumulus.

Characterized as rounded balls of semi-transparent cloud, but ill-defined and hazy. No shadows. Ci. Cu. Neb.

3. *Cirro-cumulus* (Plates 18 and 19).

Characterized as opaque rounded balls clearly defined, but showing no shadows on their under sides. Ci. Cu. Com.

Wave forms again are indicated by the addition of the word undatus.

GROUP ALTO CLOUDS. Divisible into nine species.

1. *Alto-stratus.* High stratus.

A uniform veil of cloud showing no details of structure except local variation in density in patches. Rarely dense enough to completely hide the sun, or even the full moon. A. S.

2. *Alto-stratus maculosus* (Plate 30). Mackerel sky.

Characterized as numerous nearly equal and small lenticular patches ranged on a level and about equi-distant from each other. A. S. Mac.

3. *Alto-stratus fractus* (Plate 34).

Patches and bits of cloud of irregular shape, but resembling broken bits of a level sheet. A. S. Fr.

4. *Alto-strato-cumulus* (Plate 32).

Intermediate between alto-stratus and alto-cumulus. A. S. Cu.

5. *Alto-cumulus informis* (Plate 25).

Characterized as more or less rounded cloudlets interspersed with ragged bits of cloud and occasionally with streaks of cirrus, the cloudlets showing no clear-cut outlines, but having distinct shadows. A. Cu. In.

6. *Alto-cumulus nebulosus* (Plate 26).

Hazy alto-cumulus. A. Cu. Neb.

7. *Alto-cumulus castellatus* (Plate 28). Turret cloud.

A high cloud resembling a number of tall narrow cumulus clouds on a very diminutive scale. The cloudlets show distinct shadows, are very opaque, and their upper margins are sharply defined. Vertical axes longer than the horizontal ones. A. Cu. Ca.

8. *Alto-cumulus glomeratus* (Plate 29).

Characterized by the roundness and regularity of the cloudlets, which have sharp margins, cast distinct shadows, and have their axes about equal in all directions. A. Cu. Gl.

9. *Alto-cumulus communis.*

Small high cumulus of the ordinary pyramidal pattern. A. Cu. Com.

10. *Alto-cumulus stratiformis* (Plate 27).

Flattened cloudlets gathering into small detached sheets. A. Cu. S.

Lower clouds. Group Stratus.

1. *Stratus communis* (Plates 37 and 41).

In its most typical state, stratus consists of a sheet of cloud of approximately uniform thickness. The most common form, however, does vary considerably, though usually dense enough to hide the sun. Portions of such a sheet would take the same specific name, unless the portions are very small and ragged, which would be expressed by adding the word fractus. S. Com.

2. *Stratus maculosus* (Plate 40).

Formed either by the appearance of cloud in lumps, which are always lenticular in shape, and ultimately join together to form a stratus, or by the break up of the typical stratus. S. Mac.

3. *Stratus radius* (Plate 42). Roll cloud.

Formed during the break up of a low stratus, which separates up into a number of parallel lines of cloud. S. R.

4. *Stratus lenticularis* (Plate 47). Fall cloud.

Formed by the collapse of cumulus or strato-cumulus. A cloud of evening, easily recognized as lenticular patches. S. L.

5. *Strato-cumulus* (Plates 38 and 39).

A term applied to either a stratus which has thickened every here and there into cumulus, or a number of cumulus which have joined together so as to show a nearly continuous common base. S. Cu.

GROUP CUMULUS.

1. *Cumulus minor* (Plate 43). Small cumulus.

Cumulus clouds so small as to present the appearance of rounded lumps, no definite pyramidal form or flattened base. Cu. Mi.

2. *Cumulus major* (Plates 44 and 45). Large cumulus.

Characterized by a flattened base and rounded clear-cut upper surfaces. Cu. Ma.

3. *Cumulo-nimbus* (Plates 49 to 52). Storm cloud.

Characterized by the expanded, anvil-shaped, or disc-shaped top, cirrifying at its edges.

GENERAL TERMS

Nimbus, a term applied to a cloud from which rain is falling. When the form of the cloud is visible, the term should be attached to that belonging to the cloud. It may, however, be used as a substantive alone when there is nothing to show from what sort of cloud, or combination of clouds, the rain is falling (Plates 35 and 36).

Nimbus is either heavy stratus, massive strato-cumulus, or a combination of these with stratiform clouds above, and possibly ragged masses of fracto-cumulus below. N. either alone or after the sign of the cloud.

Fracto- is a term placed as a prefix before the name of a cloud to indicate that the cloud has ragged irregular margins, as if it had been more or less torn to pieces. It is sometimes less awkward to append the word fractus after the name of the cloud.

A convenient abbreviation would be to write F. after the name of the cloud.

Undatus, or waved, should always be added to the name of any cloud which shows the arrangement so described.

CHAPTER XI
CLOUD PHOTOGRAPHY

REFERENCE has been made in the first chapter to the fact that those who wish to make a photographic study of clouds must follow a special course of procedure. For every photographic purpose there is some particular process or some special kind of apparatus which is better fitted for the end in view than any other, and half the difficulty in attaining success is to find out the best tools and the best methods.

There is no difficulty whatever in securing excellent photographs of heavy grey clouds, or of clouds which stand out dark against a twilight sky. Any camera and any plate can be used, and in an experienced hand will ensure success after a few trials, but except under these special conditions, cirrus, in all its varieties, the alto clouds, and even many of the lower ones, present a real difficulty due to two causes. In the first place, they and their surroundings are so brilliant that a very short exposure is sufficient, far shorter than would be needed for a sunlit landscape; and in the second place, the actinic value of the light they reflect is very little greater than that received from the background of blue sky. When so minute a difference comes to be represented in the monochrome of the ordinary photograph, the eye fails to appreciate it, and all the finer details are lost.

Now, if proper care is taken in the development of a negative, satisfactory results may be attained even if the exposure is twice as great, or only half as great, as it should have been to get the best result. But if the exposure is four or more times the best duration, the negative will generally yield but poor contrasts, if any result at all can be coaxed out. Again, if the exposure is only a quarter or less of the ideal time, little or no image will come out. Suppose, now, we have a brilliant object, and the correct exposure for the plate and aperture of lens employed should be one-fiftieth of a second; if we make an error either in judging or in effecting the exposure, which amounts to one twenty-fifth of a second too much, we get the negative exposed three times as much as it should be. Suppose, again, the object is less brilliant, and the correct exposure should be one-fifth of a second, an equal error of one twenty-fifth will make little difference. But in photographing cirrus and such clouds, if we used the same plates and the same lens apertures as we employ for ordinary landscape work, we should want exposures of the order of those given by a focal plane shutter, and a mistake either in judging or in making the exposure, of even the hundredth part of a second, would be fatal to good results, and would probably completely spoil the plate. Evidently one of our first steps must be to lengthen the correct exposure.

There are four ways in which this can be done—by using a slow-acting plate, by lessening the aperture of the lens, by putting some transparent screen in front of the lens to shut off some of the light, and, finally, by pointing the camera, not at the cloud itself, but at its image in a black mirror.

Of these, of course the slow plate and small aperture are the simplest to adopt, and all the cloud studies shown in the illustrations to these pages have been taken on plates prepared for photo-mechanical purposes or for transparencies. There seems to be nothing to choose between these two brands. Orthochromatic, iso-chromatic, double-coated, and many other special types of plate had previously been tried, both with coloured filters in front of the lens and without them, without showing any marked superiority over an ordinary plate of low rapidity. At last the photo-mechanical plates were tried, and the efforts made to get satisfactory cloud portraits, which had previously been marked only now and then with satisfactory results, became uniformly and continuously successful.

If the slow plates are exposed in the camera without either a screen or the black mirror, the diaphragm should be reduced to a small size and the exposure suitably adjusted. The length of exposure may generally be judged by looking at the image on the focusing screen, and reducing the aperture until the picture shows its detail easily. Then, regarding the picture as that of a sunlit sea or distant landscape, judge the necessary exposure by the brightness of the image.

No definite rule can be given. The light varies enormously from day to day, and hour to hour, and especially with the position occupied by the cloud relative to the sun. Thus, working with a lens of six inches focus and an aperture of a quarter of an inch, the exposure may vary from the quickest snap of a Thornton-Pickard roller blind to as much as a quarter of a second, or even more. Again, using a lens of eighteen inches focus and an exposure of a fiftieth of a second, the necessary aperture might vary from an eighth of an inch up to an inch and a half. But if we suppose that we are dealing with an ordinary bright summer sky between 9 a.m. and 5 p.m., and that the clouds are cirrus or cirro-cumulus, an aperture of about one thirty-second of the focal length will probably give some sort of image with a snap-shot exposure. At first the failures will be many, but a little practice will soon enable very respectable pictures to be taken by varying either the diaphragm or the speed of shutter. Heavier clouds of the alto types will need rather longer exposure or larger aperture.

The lens may be of any kind, as long as it gives a well-defined image, but there are many advantages in using one of the rectilinear type provided with an iris diaphragm. A rapid lens is not needed; indeed, it has been pointed out that slowness is a very great desideratum, and if the camera is provided with a rapid lens it must be ruthlessly stopped down. For general cloud purposes the best kind of lens is a wide-angle rectilinear, but many occasions will present themselves on which a lens of longer focus will be wanted in order to give more insight into the details of some specially delicate clouds. If the lenses are good, and the focusing is accurate, enlargements will go a long way towards revealing the minuter structures, but the results can never be quite so well defined as a direct photograph in a long camera.

A shutter will be essential, and it should be one which opens in the middle, or which travels across the lens. The shutters which are ingeniously contrived to give more exposure to the lower part of the picture than to its upper part are useless for the purpose in view. It should have some latitude of exposure, from about one-sixtieth of a second up to a full second or more.

Then as to the camera. Any light-tight camera will do, and, as the objects will all be at a great distance, it may very well be a fixed-focus one, or may be kept set up and fixed in focus for a distant object. If not, on setting it up it should be focused on the horizon or most distant object possible, and not on the cloud itself. As, however, the clouds present themselves at all heights above the horizon, even in the zenith, it becomes necessary to have some means of pointing the camera in such directions. To a certain extent the ordinary stand does allow of tilting, but a special support which will allow the camera to be fixed firmly in any position is of the greatest convenience.

If the study is meant to be at all prolonged, the best plan is to make a suitable camera, once for all, which can be left in fixed focus, so as to be always ready, and which can be directed with equal ease to any part of the sky, from the horizon to the zenith. If it is intended to use a black mirror, then a special mount becomes almost essential.

Many of the most delicate of the photographs reproduced here have been taken with a camera of peculiar pattern, the structure of which is shown in Plate 61. The lens is an ordinary rapid rectilinear, and the stop used was generally one-sixteenth of the focal length. The shutter is a light slip of aluminium, which can be drawn across from side to side at any desired pace. The body of the camera is mahogany, with a bellows part for getting correct focus, but when once this was obtained the back was clamped to the tail-board and a little varnish brushed over the clamping screws.

Plate 61.

CLOUD CAMERA FOR STUDIES.

The camera swings on a couple of screws, which act as trunnions. These pass through two upright arms, which spring on either side from the base board, which is attached to the ordinary camera stand. This base board can be rotated into any horizontal position desired, and the camera can be tilted through any vertical angle by swinging it between the uprights, and can be clamped by tightening the two trunnion screws. These screws are so placed on the front of the camera that the lens and its attachments on the one side nearly balance the back part of the camera on the other side, and so lessen the danger of slipping.

Supported in front of the lens by light brass-work is the black mirror, made of a very dark glass optically worked on the front face. It is a curious fact that, although bits of plate-glass blackened on the back seem to the naked eye to give a single image of sufficient truth, if such a mirror is placed in front of the camera the second faint image formed by reflection from the blackened surface is almost always to be detected. Moreover, the lens with its large aperture at once detects irregularities in the surface of the glass, which are quite imperceptible through the narrow limits of the pupil. Black glass, with a truly worked surface, is essential then, but the surface need not be of the high order of excellence required for mirrors used for telescopic work, since the first image is not, as a rule, intended to be highly magnified.

The mirror is held so that its surface makes an angle of about 33 degrees with the axis of the lens, and the block carrying shutter and mirror can be turned round into any position by slipping it round the lens mount as an axis. The mirror thus always retains the correct angle.

The action of the mirror is to a large extent due to mere diminution of brightness, but it also partly extinguishes the blue light of the sky without exerting any such influence on the white light from a cloud. This is due to the fact that the blue light of the sky is partly polarized, while that reflected from the cloud is not. Now, polarized light which falls upon a black mirror held in a particular position is not reflected by it. This position depends upon various circumstances, but one condition is that the reflected ray must make an angle of about 33 degrees with the surface of the glass. The amount of the polarized component of the blue light varies greatly, but is at a maximum at all points 90 degrees away from the sun. This, then, is the best possible position for photographing a cloud, as the whole of this polarized component may be suppressed by adjusting the mirror to the proper position, and then the most delicate cirrus fibres stand out brilliant on an almost black background.

The black mirror could with some advantage be replaced by a Nicol's prism mounted between the components of the lens, so that it could be turned in any position; but Nicol's prisms are expensive, and such an arrangement would cost many times the sum sufficient for an excellent mirror, and then would narrow down the field of view in a very inconvenient way.

With this apparatus exposures of a tenth to a fifth of a second were usually required for high clouds in bright daylight, while longer times, up to a second, might be required under less actively actinic conditions.

The exposure having been made, the next step is development.

Now, every practical photographer has his own pet formula, his own particular favourite among the numerous developing compounds now on the market. It is, therefore, rather a thankless task to offer advice as to which should be selected. In all probability as good results may be got by other methods and other formulæ, and the description which follows must be understood rather as an account of the process actually adopted, than advice as to that which should be chosen.

The developer used has been always pyro and ammonia, made up in accordance with the formula—

Pyro	30	grains
Potassium metabisulphite	30	„
Ammonium bromide	30	„
Water	10	ozs.

But if much work was anticipated the solution was made up in a more concentrated form, and diluted to this strength of 3 grains of pyro per ounce for actual use.

The ammonia solution is prepared by mixing 3 drams ammonia fortiss. with 20 ozs. of water.

In developing it is necessary to remember that our object is to make the most of a very small difference in effect. The plate is first flowed over with a mixture of sufficient developer, with not more than a quarter of its bulk of the ammonia. If the cloud should flash out in a few seconds add more of the pyro solution, but unless the exposure has been much overdone this will not happen. If the image begins to appear after from thirty to forty seconds it is probable that the best result will be reached by leaving it alone, but if there is any hanging back of the detail another quarter bulk of ammonia should be put into the glass, the developer mixed with it, and the whole returned to the developing dish.

If no image appears after about forty seconds, add more ammonia as above described, and leave for another forty seconds, and so on, until by this method of trial the right quantity of alkali for the particular exposure has been ascertained. The development must never be hurried, or the background of sky will blacken too soon, and in some cases it may take a quarter of an hour or more to get enough density on the cloud. But as a general rule the image is fully out in about two minutes, and the plate is then washed and fixed in the usual way.

If a black mirror is used there will seldom be any necessity for intensification, but if not, it may frequently be required, especially for the more delicate kinds of cirrus. Indeed, the image may sometimes be so thin that the common process of intensification by mercury and ammonia does not give density enough. If that seems at all likely to be the case, it is wiser to use the formula known as Monckhoven's, since that simply adds silver to silver instead of replacing the silver image by some other body, and the process can consequently be repeated more than once, if sufficient density is not secured by the first application. The formula does not seem to be very often used, so it may be best to quote it.

A.	Potassium bromide	100	grains
	Mercuric chloride	100	„
	Water	10	ozs.

B.	Potassium cyanide (pure)	100	grains
	Silver nitrate	100	„
	Water	10	ozs.

Place the washed negative in A until it has gone white, then rinse it well and transfer to B, in which the image turns to a velvety black. After washing, the process can be repeated.

Intensification is, however, only a way of saving photographs which cannot be secured again. If the first photograph of a particular variety of cloud is not satisfactory, it ought at least to tell the operator where he had gone wrong, and a second attempt should produce a better result than any image built up by chemical action on an imperfect base.

There is nothing novel in any of these methods, and there is no doubt that other formulæ would be as good; but the one thing essential is to have a developer whose action can be held under control, and to apply that developer in such a way that very considerable over-exposure will not result in the ruin of the plate. If a number of photographs have been taken in about the same part of the sky, and within a short time of each other, then the correct proportions of developer and alkali will be nearly the same for all, but the first of such a batch will always have to be attacked in the cautious step-by-step method. Patience and perseverance, backed by a steady refusal to be discouraged by the failures which are at first inevitable, are as certain to be crowned by success as they are in other studies.

The workers are few, and there is much to be done; for it is mainly to those who will photograph the higher clouds, and so trace the stages of their growth and decay, that we must look for the data which will enable us to solve the problems they present, and so enlarge the narrow boundaries of our knowledge of some of the most beautiful things in Nature.

REFERENCES

1. "International Atlas of Clouds" (Atlas International des Nuages). Hildebrandsson, Riggenbach, and Teisserenc de Bort. Paris. 1896.

This is the atlas referred to in the text. The letter-press is short, and is repeated in English, French, and German.

2. "Annals of the Astronomical Observatory of Harvard College." Vol. XXX. Observations made at the Blue Hill Meteorological Observatory. Part III. Measurement of Cloud Heights and Velocities. By H. H. Clayton and S. P. Fergusson. Part IV. Discussion of the Cloud Observations. By H. H. Clayton.

This last gives a very concise account of all the different proposals which have been made for the systematic naming of clouds.

3. "Études International des Nuages." 1896-1897. Observations et Mesures de la Suède. I., II. Publication de l'Observatoire Météorologique de l'Université Roy. d'Upsala. H. H. Hildebrandsson.

An account of the Upsala observations referred to in the text.

4. *Quarterly Journal of the Royal Meteorological Society.*

Helm Wind. Marriott. 1886 and 1889.

The Thickness of Shower Clouds. Clayden. 1886.

Methods of Cloud Measurement. Ekholm. 1888.

Cirrus Formation. Clayton. 1890.

Nomenclature of Clouds. Hildebrandsson. 1887.

        " Abercromby. 1887.

        " Wilson-Barker. 1890.

        " Gaster. 1893.

        " Scott. 1895.

A New Instrument for Cloud Measurement. Ekholm. 1893.

Calculation of Photographic Cloud Measurements. Olsson. 1894.

The Motion of Clouds. Shaw. 1895.

5. Reports of the British Association. Reports of the Committee on Meteorological Photography. Clayden. 1891 to 1900.

The reports for 1896 and 1900 refer mainly to the measurements described in the text.

6. "Cloudland." Clement Ley.

The work in which Mr. Ley set forth his proposed scheme.

7. "A Popular Treatise on the Winds." Ferrel.

Not a "popular" work in the usual sense, but contains lucid descriptions of the mechanics of the atmosphere.

8. There are many excellent text-books on meteorology, all of which deal more or less with the movements of the atmosphere and the formation of clouds.

THE END

INDEX

Lector House believes that a society develops through a two-fold approach of continuous learning and adaptation, which is derived from the study of classic literary works spread across the historic timeline of literature records. Therefore, we aim at reviving, repairing and redeveloping all those inaccessible or damaged but historically as well as culturally important literature across subjects so that the future generations may have an opportunity to study and learn from past works to embark upon a journey of creating a better future.

This book is a result of an effort made by Lector House towards making a contribution to the preservation and repair of original ancient works which might hold historical significance to the approach of continuous learning across subjects.

HAPPY READING & LEARNING!

LECTOR HOUSE LLP
E-MAIL: lectorpublishing@gmail.com